AF597913

METHODS IN MOLECULAR BIOLOGY™

Series Editor
John M. Walker
School of Life Sciences
University of Hertfordshire
Hatfield, Hertfordshire, AL10 9AB, UK

For further volumes:
http://www.springer.com/series/7651

PIWI-Interacting RNAs

Methods and Protocols

Edited by

Mikiko C. Siomi

Graduate School of Science, The University of Tokyo, Tokyo, Japan

Editor
Mikiko C. Siomi
Graduate School of Science
The University of Tokyo
Tokyo, Japan

ISSN 1064-3745 ISSN 1940-6029 (electronic)
ISBN 978-1-62703-693-1 ISBN 978-1-62703-694-8 (eBook)
DOI 10.1007/978-1-62703-694-8
Springer New York Heidelberg Dordrecht London

Library of Congress Control Number: 2013950180

Printed on acid-free paper

Humana Press is a brand of Springer
Springer is part of Springer Science+Business Media (www.springer.com)

Preface

The study of small noncoding RNAs was accelerated by the discovery of RNA interference (RNAi) in the nematode *Caenorhabditis elegans* in 1998, by two Nobel Prize laureates Andrew Fire and Craig Mello. Researchers have subsequently focused on understanding how short interfering RNAs (siRNAs) and microRNAs (miRNAs), which represent two major RNAi-triggering small noncoding RNAs, actually induce gene silencing and what mechanisms are responsible for producing them from double-stranded RNA precursors. Studies revealed that RNAi is highly conserved in a wide variety of animals and plants, as well as in unicellular organisms such as fission yeast. They further demonstrated that in addition to *C. elegans*, other model species such as *Drosophila melanogaster*, *Arabidopsis thaliana*, and mice, along with cultured cell lines such as HeLa and *Drosophila* Schneider 2 (S2) could serve as useful tools for investigating the molecular nature of RNAi. Indeed, these organisms have been used to address the fundamental questions of the biogenesis and functions of RNAi-triggering small noncoding RNAs. It is clear that the ubiquity of RNAi has a significant impact on our understanding of the RNAi world.

PIWI-interacting RNAs (piRNAs) are the third and most-recently discovered group of silencing-inducible small RNAs in animals. The delay in their discovery was mainly the result of their expression predominantly in the gonads. Cultured HeLa and S2 cells, which are otherwise excellent tools for studying ubiquitous RNAi, do not express piRNAs to a detectable level. Although the limited expression of these piRNAs has meant they have been less well studied than other small RNAs, it has conversely stimulated researchers' curiosities. Comprehensive high-throughput sequencing analysis of piRNAs in embryos, testes, and ovaries of *D. melanogaster*, as well as in mouse and rat testes, has raised the profile of piRNAs and thus further accelerated piRNA studies.

piRNAs associate with specific members of the Argonaute family of proteins to form RNA-induced silencing complexes (RISCs), as with siRNAs and miRNAs, and piRNA-containing RISCs (piRISCs) implement RNA silencing similarly to siRISCs and miRISCs, though their mechanisms may vary. This indicates the equivalence of the various small RNAs. However, piRNA precursors are thought to be single-stranded because piRNA processing occurs independently of Dicer. piRNAs are several nucleotides longer than miRNAs and siRNAs. Most piRNAs are complementary to transcripts of transposable elements and represses the mobile genomic elements to maintain the integrity of the genome in the germline. These piRNA-specific features further highlight the uniqueness of piRNAs, prompting studies aimed at developing a comprehensive understanding of the small RNA world. However, the process whereby piRNAs mature from their precursors remains unknown. We still do not understand how piRNA precursors are selectively chosen or the mechanisms responsible for silencing of target elements. Given these unanswered basic questions, it is clear that the piRNA studies are still a long way from their final goal. In order to further progress towards this goal, this is an ideal time to gather and share our expertise and knowledge on the experimental methods used to unveil the piRNA world. This book "PIWI-interacting RNAs: Methods and Protocols" provides the most recent methods and protocols for studying piRNAs in the gonads of a wide range of species, as well as in any other organs where piRNAs may be detected. This book will help both established researchers and newcomers to the field to progress towards the ultimate goal of understanding the mechanisms and actions of piRNAs.

Tokyo, Japan **_Mikiko C. Siomi_**

Contents

Contributors

ALEXEI A. ARAVIN • *Division of Biology, California Institute of Technology, Pasadena, CA, USA*
SÉVERINE CHAMBEYRON • *Institut de Génétique Humaine, CNRS, Montpellier, Cedex, France*
DAHUA CHEN • *State Key Laboratory of Reproductive Biology, Institute of Zoology, Chinese Academy of Sciences, Beijing, China*
GUAN-WEI CHIRN • *Department of Biology, Rosenstiel Basic Medical Science Research Center, Brandeis University, Waltham, MA, USA*
BENJAMIN CZECH • *Cold Spring Harbor Laboratory, Cold Spring Harbor, NY, USA*
THOMAS GRENTZINGER • *Institut de Génétique Humaine, CNRS, Montpellier, France*
SHOZO HONDA • *Department of Biomedical Sciences, Cedars-Sinai Medical Center, Samuel Oschin Comprehensive Cancer Institute, Los Angeles, CA, USA*
HAIDONG HUANG • *State Key Laboratory of Reproductive Biology, Institute of Zoology, Chinese Academy of Sciences, Beijing, China*
HSIN-YI HUANG • *Ontario Cancer Institute, Princess Margaret Hospital, Toronto, ON, Canada*
GABOR L. IGLOI • *Institute of Biology, University of Freiburg, Freiburg, Germany*
DAISUKE ITO • *Department of Pathology, Medical School, Osaka University, Osaka, Japan*
ALLA KALMYKOVA • *Institute of Molecular Genetics of Russian Academy of Sciences, Moscow, Russia*
SUSUMU KATSUMA • *Department of Agricultural and Environmental Biology, Graduate School of Agricultural and Life Sciences, The University of Tokyo, Tokyo, Japan*
SHINPEI KAWAOKA • *Cold Spring Harbor Laboratory, Cold Spring Harbor, NY, USA*
RENÉ F. KETTING • *Institute of Molecular Biology (IMB), Faculty of Biology, University of Mainz, Mainz, Germany*
YOHEI KIRINO • *Computational Medicine Center, Department of Biochemistry and Molecular Biology, Thomas Jefferson University, Philadelphia, PA, USA*
YORIKO KIRINO • *Department of Biomedical Sciences, Cedars-Sinai Medical Center, Samuel Oschin Comprehensive Cancer Institute, Los Angeles, CA, USA*
KANAKO KITA-KOJIMA • *Department of Pathology, Medical School, Osaka University, Osaka, Japan*
REINA KOMIYA • *Experimental Farm, National Institute of Genetics (NIG), Mishima, Shizuoka, Japan*
HIROTAKA KOSHIMA • *Department of Pathology, Medical School, Osaka University, Osaka, Japan*
SATOMI KURAMOCHI-MIYAGAWA • *Department of Pathology, Medical School, Osaka University, Osaka, Japan*
HENRIETTE M. KURTH • *Viollier AG Spalenring, Basel, Switzerland*
NELSON C. LAU • *Department of Biology, Rosenstiel Basic Medical Science Research Center, Brandeis University, Waltham, MA, USA*

SERGEY LAVROV • *Institute of Molecular Genetics of Russian Academy of Sciences, Kurchatov, Moscow, Russia*
ADRIEN LE THOMAS • *Division of Biology, California Institute of Technology, Pasadena, CA, USA*
HAIFAN LIN • *Department of Cell Biology, Yale Stem Cell Center, Yale School of Medicine, New Haven, CT, USA*
JESSICA A. MATTS • *Department of Biology, Rosenstiel Basic Medical Science Research Center, Brandeis University, Waltham, MA, USA*
JON MCGINN • *Cold Spring Harbor Laboratory, Cold Spring Harbor, NY, USA*
KENJYO MIYAUCHI • *Department of Chemistry and Biotechnology, Graduate School of Engineering, University of Tokyo, Tokyo, Japan*
KAZUFUMI MOCHIZUKI • *Institute of Molecular Biotechnology of the Austrian Academy of Sciences, Vienna, Austria*
ZISSIMOS MOURELATOS • *Department of Pathology and Laboratory Medicine, Division of Neuropathology, University of Pennsylvania School of Medicine, Philadelphia, PA, USA*
TORU NAKANO • *Department of Pathology, Medical School, Osaka University, Osaka, Japan*
YUZO NIKI • *Department of Biology, Faculty of Science, Ibaraki University, Mito, Ibaraki, Japan*
KEN-ICHI NONOMURA • *Experimental Farm, National Institute of Genetics (NIG), Mishima, Shizuoka, Japan*
TOMOKO NOTO • *Institute of Molecular Biotechnology of the Austrian Academy of Sciences, Vienna, Austria*
IVAN OLOVNIKOV • *Division of Biology, California Institute of Technology, Pasadena, CA, USA*
RAMESH S. PILLAI • *EMBL Grenoble Outstation, Grenoble, Cedex, France*
MICHAEL REUTER • *EMBL Grenoble Outstation, Grenoble, Cedex, France*
AYAKA SAISHO • *Department of Biology, Faculty of Science, Ibaraki University, Mito, Ibaraki, Japan*
KUNIAKI SAITO • *Department of Molecular Biology, Keio University School of Medicine, Tokyo, Japan*
YURIKO SAKAGUCHI • *Department of Chemistry and Biotechnology, Graduate School of Engineering, University of Tokyo, Tokyo, Japan*
TAKUYA SATO • *Department of Biology, Faculty of Science, Ibaraki University, Mito, Ibaraki, Japan*
YUSUKE SHIROMOTO • *Department of Pathology, Medical School, Osaka University, Osaka, Japan*
SERGEY SHPIZ • *Institute of Molecular Genetics of Russian Academy of Sciences, Kurchatov, Moscow, Russia*
TAKEO SUZUKI • *Department of Chemistry and Biotechnology, Graduate School of Engineering, University of Tokyo, Tokyo, Japan*
TSUTOMU SUZUKI • *Department of Chemistry and Biotechnology, Graduate School of Engineering, University of Tokyo, Tokyo, Japan*
YULIYA SYTNIKOVA • *Department of Biology, Rosenstiel Basic Medical Science Research Center, Brandeis University, Waltham, MA, USA*
YUKIHIDE TOMARI • *Institute of Molecular and Cellular Biosciences, The University of Tokyo, Tokyo, Japan*

Hiroshi Uetake • *Department of Biology, Faculty of Science, Ibaraki University, Mito, Ibaraki, Japan*

Anastassios Vourekas • *Department of Pathology and Laboratory Medicine, Division of Neuropathology, University of Pennsylvania School of Medicine, Philadelphia, PA, USA*

Hailong Wang • *State Key Laboratory of Reproductive Biology, Institute of Zoology, Chinese Academy of Sciences, Beijing, China*

Hidenori Watanabe • *Department of Biology, Faculty of Science, Ibaraki University, Mito, Ibaraki, Japan*

Takafumi Yamaguchi • *Department of Biology, Faculty of Science, Ibaraki University, Mito, Ibaraki, Japan*

Hang Yin • *Ottawa Hospital Research Institute, Ottawa, ON, Canada*

Chapter 1

Chromatin Immunoprecipitation Assay of Piwi in *Drosophila*

Hang Yin and Haifan Lin

Abstract

The generation of high-resolution maps of the epigenome is crucial to research in epigenetics, developmental biology, and stem cell biology. In recent years, small RNA pathways have been implicated in epigenetic regulation. All small RNA pathways involve Argonaute proteins as their key biogenesis and effector components. In this chapter, we describe a chromatin immunoprecipitation method for whole-genome mapping of *Drosophila* Piwi, the defining member of the Argonaute protein family. This method should have general utility for mapping other chromatin-associated factors.

Key words Piwi, Chromatin immunoprecipitation assay, Epigenetics

1 Introduction

Drosophila Piwi is the founding member of the evolutionarily conserved Argonaute/Piwi protein family, which is divided into Argonaute (Ago) and Piwi subfamilies. The Ago proteins associate with small-interfering RNAs (siRNAs) and microRNAs (miRNAs) of 20–23 nucleotides (nt), whereas the Piwi proteins interact with Piwi-interacting RNAs (piRNAs) of mostly 24–32 nt [1].

In *Drosophila*, the Piwi subfamily proteins consist of Piwi, Aubergine, and Ago3. Besides its well-known function in posttranscriptional silencing of transposons, Piwi and some of its mammalian homologs also play an important role in epigenetic regulation. First, Piwi is a nuclear protein that directly binds to chromatin [2, 3]. Second, Piwi is a typical suppressor of position effect variegation [2], similar to other key epigenetic factors such as heterochromatin protein 1 (HP1a). Third, Piwi directly interacts with other key epigenetic factors such as HP1a [4]. Fourth, *piwi* deficiency results in global loss of methylation of histone 3 at lysine 9 (H3K9me) and the delocalization of HP1 from polytene chromosomes [2] and loss

Mikiko C. Siomi (ed.), *PIWI-Interacting RNAs: Methods and Protocols*, Methods in Molecular Biology, vol. 1093,
DOI 10.1007/978-1-62703-694-8_1, © Springer Science+Business Media, LLC 2014

of euchromatic features at a subtelomeric region of chromosome 3R in adult flies [3]. Thus, Piwi and its homologs may behave as key components of an epigenetic regulatory complex required for euchromatin/heterochromatin assembly.

Chromatin immunoprecipitation (ChIP) is a widely used method for mapping the interactions of epigenetic factors with DNA [5]. Briefly, DNA and its associated proteins are covalently crosslinked in vivo and the crosslinked chromatin is then randomly sheared into short fragments. The DNA fragments associated with the protein(s) of interest are selectively immunoprecipitated and enriched. After reverse crosslinking, the associated DNA fragments are purified and their relative enrichment (comparing to the whole genome) can be determined by quantitative PCR or hybridization to genome tiling arrays or alternatively their sequences can be determined by high-throughput sequencing. The input material for a ChIP experiment can be immortalized cell lines, primary cells, tissues/organs, or even whole organisms.

Piwi is expressed in germ plasm, primordial germ cells (PGCs) [6], and adult germ cells (including germline stem cells, GSCs) [7]. Besides, *Drosophila* Piwi is also present in some somatic cells, such as in larval imaginal disc cells [8], in larval salivary gland cells [2] as well as in terminal filament cells and cap cells of the adult female gonad [7]. The presence of Piwi, Aubergine, and Ago3 proteins in *Drosophila* cell lines has not yet been systematically characterized. Piwi proteins are reportedly not expressed in *Drosophila* Schneider 2 (S2) cells [9], which was originally derived from a primary culture of late stage (20–24 h) embryos [10]. While future investigation on expression of Piwi proteins in various *Drosophila* cell lines and establishment of primary cultures for GSCs are warranted, we developed a protocol for immunoprecipitation of Piwi-bound chromatin (Piwi-ChIP) from crosslinked nuclei isolated from adult whole flies (Fig. 1). We reason that such a Piwi-ChIP using adult whole fly as input would generate similar results to that derived from dissected gonads, given the abundant presence of Piwi protein in the adult gonadal cells and the almost undetectable presence of Piwi outside the gonad. The presence of extra chromatin fragments that originate from non-gonadal tissues would be no difference as to the chromatin fragments that are derived from gonadal tissues but not bound by Piwi, and hence would not affect the enrichment of Piwi-bound chromatin by anti-Piwi antibody. By combining this Piwi-ChIP method with high-throughput sequencing, we successfully mapped the genome-wide localization of chromatin-bound Piwi protein across the *Drosophila* genome at high resolution [11].

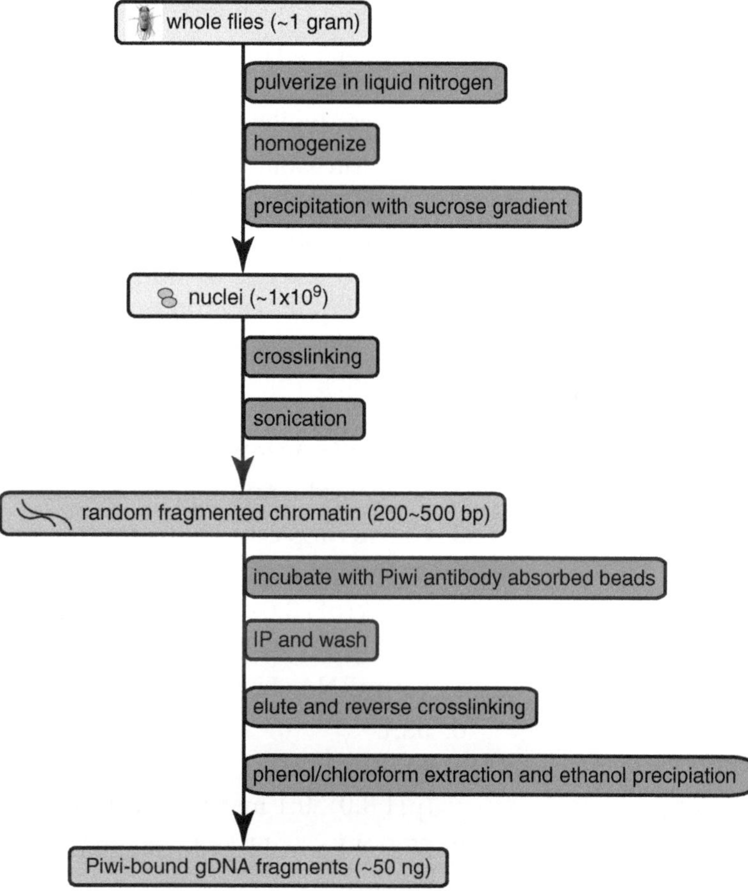

Fig. 1 Flow chart of chromatin immunoprecipitation assay of Piwi in *Drosophila*. For details, see text

2 Materials

2.1 Isolation of Nuclei

The nuclear isolation protocol is based on Shaffer et al [12].

1. Wild type *Drosophila melanogaster* adult flies (e.g., *Canton-S*).
2. CO_2 for anesthetizing flies.
3. Liquid nitrogen.
4. Mortar and pestle (0.1-0.5 ml capacity).
5. 7 mL Dounce glass tissue grinder homogenizer (Thomas scientific).
6. Buffer A+: 60 mM KCl, 15 mM NaCl, 13 mM EDTA (pH 8.0), 0.1 mM EGTA, 15 mM HEPES (pH 7.4), 0.5 mM DTT, 0.5 % NP-40, 1× Protease Inhibitor Cocktail.
7. 15 mL Wheaton glass tissue grinder homogenizer set (Thomas scientific).

8. Miracloth (Millipore).
9. Buffer AS: 60 mM KCl, 15 mM NaCl, 1 mM EDTA (pH 8.0), 0.1 mM EGTA, 15 mM HEPES (pH 7.4), 0.3 M sucrose, 1× Protease Inhibitor Cocktail.
10. Beckman Coulter Optima L-100 XP Ultracentrifuge with a SW-40 ti rotor.
11. Buffer A: 60 mM KCl, 15 mM NaCl, 1 mM EDTA (pH 8.0), 0.1 mM EGTA, 15 mM HEPES (pH 7.4), 0.5 mM DTT, 1 mM PMSF, 1× Protease Inhibitor Cocktail.
12. 7 mL Wheaton glass tissue grinder homogenizer set (Thomas scientific).

2.2 Piwi Chromatin Immunoprecipitation

1. Protein G Agarose, fast flow (Millipore).
2. Siliconized 2 mL Eppendorf tubes.
3. Counter-top centrifuge with refrigeration function.
4. ChIP Buffer: 75 mM NaCl, 50 mM HEPES (pH 7.4), 1 mM EDTA (pH 8.0), 1 mM DTT, 5 mM $MgCl_2$, 0.2 % Triton-X100, 1× Protease Inhibitor Cocktail.
5. Yeast tRNA (Life Technologies).
6. BSA.
7. Resuspension Buffer: 60 mM KCl, 15 mM NaCl, 1 mM EDTA (pH 8.0), 0.1 mM EGTA, 15 mM HEPES (pH 7.4).
8. 37.5 % Formaldehyde (Sigma).
9. Glycine.
10. ChIP Lysis Buffer: 75 mM NaCl, 50 mM HEPES (pH 7.4), 1 mM EDTA, 1 mM DTT, 5 mM $MgCl_2$, 1 % Triton-X100, 1× Protease Inhibitor Cocktail.
11. Water bath-based sonication device, e.g., Diagenode Bioruptor.
12. Siliconized 15 mL Falcon tubes.
13. ChIP Dilution Buffer: 75 mM NaCl, 50 mM HEPES (pH 7.4), 1 mM EDTA, 1 mM DTT, 5 mM $MgCl_2$, 1× Protease Inhibitor Cocktail.
14. Anti-Piwi antibody raised in guinea pigs against a Piwi C-terminal peptide (raised in Lin Laboratory).
15. RIPA 150 Buffer: 50 mM HEPES (pH 7.4), 150 mM NaCl, 2 mM EDTA (pH 8.0), 0.1 % Triton-X100, 0.1 % SDS.
16. RIPA 500 Buffer: 50 mM HEPES (pH 7.4), 500 mM NaCl, 2 mM EDTA (pH 8.0), 0.1 % Triton-X100, 0.1 % SDS.
17. LiCl Buffer: 50 mM HEPES (pH 7.4), 0.25 M LiCl, 1 mM EDTA (pH 8.0), 0.1 % NP-40.
18. TE Buffer: 10 mM Tris–HCl (pH 8.0), 1 mM EDTA (pH 8.0).

19. Elution Buffer: 0.1 M $NaHCO_3$, 1 % SDS.
20. 20 mg/mL Proteinase K (New England Biolabs).
21. NaCl.
22. Heat block (set at 65 °C).
23. DNase-free RNase A (Sigma).
24. PhaseLocking gels heavy 1.5 mL (5 prime).
25. Phenol:Chloroform:Isoamyl Alcohol (25:24:1, v/v) (Fisher).
26. 15 mg/mL Glycoblue (Life Technologies).
27. 100 % Ethanol (−20 °C).
28. 70 % Ethanol (−20 °C).

3 Methods

3.1 Isolation of Nuclei

The nuclear isolation protocol is based on Shaffer et al [12].

1. Collect about 1 g adult flies by CO_2 anesthesia.
2. Transfer flies (~5 mL) into 50 mL Falcon tubes and snap freeze in liquid nitrogen.
3. Transfer frozen flies into a porcelain mortar (~0.5 L capacity).
4. Slowly pour liquid nitrogen into the mortar. Grind with pestle while liquid nitrogen evaporates.
5. Repeat **step 4** until the adult flies have been pulverized into fine powder (*see* **Note 1**).
6. Transfer the fine powder into a 7 mL Dounce glass tissue grinder homogenizer (precooled on ice) by a metal weighting spoon.
7. Add 5 mL ice-cold Buffer A+ into the 7 mL Dounce homogenizer, briefly resuspend the powder by pipetting.
8. Disrupt the powder slurry with 10 strokes on ice. We advise use a homogenizer with a loose clearance of 0.0889-0.127mm.
9. Transfer the powder slurry into a 15 mL Wheaton glass tissue grinder homogenizer (precooled on ice) by pipetting. We advise use a homogenizer with a loose clearance of 0.0254–0.076mm.
10. Disrupt the powder slurry with 20 strokes (with the tight pestle) on ice (*see* **Note 2**).
11. Filter the crude homogenate through double-layered Miracloth (*see* **Note 3**).
12. Add 2 mL ice-cold Buffer AS into a polyethylene tube fit with Beckman SW-40 ti rotor.
13. Carefully load the filtered crude homogenate onto the 2 mL Buffer AS by a Pasteur pipette (*see* **Note 4**).

14. Carefully insert the tube into a SW-40ti rotor adaptor and centrifuge at 3,500 rpm (~2,175 × *g*) for 12 min at 4 °C on a Beckman Coulter Optima L-100 XP Ultracentrifuge (*see* **Note 5**).
15. Slowly remove the top layer (in yellow color, containing the cytosolic fraction) and the Buffer AS layer sequentially by a Pasteur pipette.
16. Resuspend the crude nuclei pellet (in creamy white or light grey color) in 5 mL Buffer A.
17. Transfer the crude nuclei suspension into a 7 mL Wheaton glass tissue grinder homogenizer (precooled on ice).
18. Disrupt the crude nuclei suspension with 10 strokes (the "tight" pestle).
19. Repeat **steps 13–15** for another time.
20. Stain nuclei by DAPI or Hoechst 33258 and count under a Fluorescence microscope (Checkpoint #1, *see* **Note 6**).
21. Aliquot the nuclei in 2 mL siliconized eppendorf tubes.

3.2 Piwi Chromatin Immunoprecipitation

1. Transfer Protein G Agarose beads slurry into two 2 mL siliconized eppendorf tubes (500 μL slurry in each tube).
2. Centrifuge at 100 × *g* for 30 s at 4 °C in a counter-top centrifuge.
3. Remove the supernatant and resuspend each bead pellet with 1 mL ChIP Buffer (precooled on ice).
4. Repeat **steps 2–3** for two more times.
5. After the last centrifugation, resuspend each bead pellet with 1 mL ChIP Buffer (precooled on ice) supplemented with 200 μg/mL tRNA and 1 % BSA. Equilibrate the beads by rotating the tubes at 4 °C for 1 h.
6. Resuspend 1×10^9 nuclei into 1 mL Resuspension Buffer (precooled on ice) by repeated pipetting. Avoid any nuclei clump (*see* **Note 7**).
7. Add formaldehyde to a final concentration of 0.2 % and crosslink at room temperature for 15 min with constant agitation (*see* **Note 8**).
8. Add glycine to a final concentration of 0.125 M and quench the crosslinking at room temperature for 5 min with constant agitation.
9. Centrifuge in a counter-top centrifuge at 4 °C for 2 min at maximal speed to collect the crosslinked nuclei at the bottom. Aspirate the supernatant (*see* **Note 9**).
10. Resuspend the crosslinked nuclei into 2 volumes (~300 μL) of ChIP Lysis Buffer.

11. Sonicate in a Diagenode Bioruptor (high power setting) with circulating cold water for 15 cycles with 30 s on and 60 s off (Checkpoint #2, *see* **Note 10** for how to optimize sonication condition).
12. Centrifuge at 12,000 × *g* for 10 min at 4 °C in a counter-top centrifuge.
13. Transfer the supernatant to a 15 mL siliconized Falcon tube. Avoid carrying over any pellet.
14. Dilute the supernatant with 4 volumes of ChIP Dilution Buffer (*see* **Note 11**).
15. Add the equilibrated beads from **step 5** (one tube) to the diluted supernatant. Preclear the nuclear lysate by rotating at 4 °C for 1 h.
16. Add 20 μg anti-Piwi antibody (generated in Lin laboratory, raised in guinea pigs against a Piwi C-terminal peptide, Protein A/G affinity purified) to the equilibrated beads in another tube. Bind the antibody on the beads by rotating at 4 °C for 1 h (*see* **Note 12**).
17. Centrifuge the precleared nuclear lysate (with beads) at 100 × *g* for 30 s.
18. Transfer the precleared nuclear lysate to a new 15 mL siliconized Falcon tube. Aliquot ~1/100 volume of precleared nuclear lysate in a 2 mL siliconized eppendorf tube (labeled with "Input lysate") and store at −20 °C.
19. Add antibody-bound beads from **step 16** to the remaining precleared nuclear lysate. Rotate at 4 °C overnight.
20. Aliquot ~1/100 volume of nuclear lysate with beads into a 2 mL siliconized eppendorf tube (labeled with "pre-wash beads"). Centrifuge the tube at 100 × *g* for 30 s. Aspirate the supernatant and store at −20 °C.
21. Centrifuge the remaining nuclear lysate with beads at 1,000 rpm for 30 s.
22. Aliquot ~1/100 volume of supernatant in a 2 mL siliconized eppendorf tube (labeled with "post-ChIP lysate") and store at −20 °C.
23. Aspirate the remaining supernatant.
24. Add 10 mL RIPA 150 Buffer to the bead pellets. Rotate at 4 °C for 20 min.
25. Centrifuge the bead slurry at 100 × *g* for 30 s. Aspirate the supernatant.
26. Add 10 mL RIPA 500 Buffer to the bead pellets. Rotate at 4 °C for 20 min.
27. Centrifuge the bead slurry at 100 × *g* for 30 s. Aspirate the supernatant.

28. Add 10 mL LiCl Buffer to the bead pellets. Rotate at 4 °C for 20 min.
29. Centrifuge the bead slurry at 100×*g* for 30 s. Aspirate the supernatant.
30. Add 10 mL TE Buffer to the bead pellets. Rotate at 4 °C for 20 min.
31. Aliquot ~1/100 volume of bead slurry in TE Buffer into a 2 mL siliconized eppendorf tube (labeled with "post-wash beads"). Centrifuge at 100×*g* for 30 s. Aspirate the supernatant and store at −20 °C.
32. Centrifuge the remaining bead slurry at 100×*g* for 30 s. Aspirate the supernatant (Checkpoint #3, *see* **Note 13**).
33. Add 500 μL Elution Buffer to the bead pellet. Rotate at room temperature for 30 min (*see* **Note 14**).
34. Centrifuge the bead slurry at 100×*g* for 30 s. Transfer the supernatant to a 2 mL siliconized eppendorf tube.
35. Add another 500 μL Elution Buffer to the bead pellet. Rotate at room temperature for 30 min.
36. Centrifuge the bead slurry at 100×*g* for 30 s. Combine the supernatant with the previous supernatant in the 2 mL siliconized eppendorf tube.
37. Add 8 μL Proteinase K (20 mg/ml) to the 1 mL combined eluate and incubate at 37 °C for 30 min.
38. Add NaCl to a final concentration of 0.3 M to the eluate.
39. Transfer the eluate to a heat block set at 65 °C. Incubate for at least 6 h to reversed crosslink the sample.
40. Add 2 μL DNase-free RNase A to the eluate and incubate at the room temperature for 10 min.
41. Centrifuge two PhaseLocking gels heavy tubes for 2 min at maximal speed. Transfer and split the eluate into two PhaseLocking gels (~500 μL each).
42. Add 500 μL phenol/chloroform/isoamyl alcohol. Shake vigorously for 20 s (*see* **Note 15**).
43. Centrifuge the sample at 10,000×*g* for 5 min at 4 °C in a counter-top centrifuge.
44. Transfer the upper aqueous phase (above the gel) into two 1.5 mL siliconized Eppendorf tubes.
45. Add 1 mL GlycoBlue (15 mg/mL) to each tube and mix well. Add 2 volumes of anhydrous Ethanol (precooled at −20 °C) to each tube and mix well.
46. Incubate the tubes at −80 °C for at least 1 h or at −20 °C for overnight.

47. Centrifuge the tubes at 20,000 × *g* for 15 min at 4 °C in a counter-top centrifuge.
48. Carefully remove the supernatant by pipetting.
49. Wash the blue pellets twice with 1 mL 70 % Ethanol (precooled at −20 °C).
50. Carefully remove the supernatant by pipetting. Air dry the pellet for 30 min or until the Ethanol evaporates off.
51. Dissolve the pellets in 20 μL of TE Buffer (*see* **Note 16**).

4 Notes

1. Pour liquid nitrogen half full and grind slowly to avoid spill-over. Stop grinding when liquid nitrogen is almost evaporated in order to collect powder at the very bottom of the mortar.
2. It might be difficult for the beginning several strokes. Twist the pestle to help the slurry go through the clearance. If strokes are too difficult, grind powder in liquid nitrogen more thoroughly next time.
3. Wet Miracloth with Buffer A+ to avoid unnecessary absorbance of the crude homogenate on Miracloth.
4. Start the loading by putting the tip of Pasteur pipette close to the upper surface of the Buffer AS. Try to load slowly to avoid disturbance of the Buffer AS cushion.
5. Precool the centrifuge at 4 °C. The rotor should be well-balanced before centrifuge.
6. Checkpoint #1: usually, $0.5\text{–}1 \times 10^9$ nuclei can be isolated from 1 g adult flies. If strokes in **step 10** are too difficult to complete, one can start with 0.5 g of adult flies. If necessary, nuclei can be stored in Buffer A supplemented with 50 % Glycerol at −80 °C for several days.
7. If nuclei are frozen at −80 °C, thaw nuclei on ice and wash the thawed nuclei once with 10 volumes of Resuspension Buffer.
8. Formaldehyde solution is toxic. Perform this step in a fume hood.
9. Nuclei tend to form clumps after crosslinking. Dispense the nuclei clump by passing the nuclei suspension through a 36-gauge needle.
10. Checkpoint #2: the number of sonication cycles provided here is only as a reference. The optimal sonication cycle number should be determined empirically for each sonicator. To do this, save 1/20 of sample before sonication (cycle 0) and take out 1/20 of sample from the tube at sonication cycle number 5, 9, 11, 13, 15, 17, 19, 21, 25. Replenish the tube with same

volume of ChIP Lysis Buffer to keep the volume in the tube constant. Follow **steps 37–50** for all sonication samples. Load all sonication samples side by side onto a 1.5 % Agarose/TAE gel with 100-bp DNA ladder. Determine the sonication cycle number that gives rise to a DNA smear mainly migrate between 200 and 500 bp.

11. The 1 % Triton-X100 in the ChIP Lysis Buffer will be diluted to 0.2 % by the ChIP Dilution Buffer. Antibody recognition to its epitope is usually not affected by 0.2 % Triton-X100.
12. Other commercially available anti-Piwi antibodies need to be tested empirically for their usage with this protocol (*see* **Note 13**). Always perform a control ChIP with a nonspecific antibody raised in the same species as the anti-Piwi antibody.
13. Checkpoint #3: a quality control should be performed at this step to confirm the immunoprecipitation of Piwi protein by anti-Piwi antibody. To do this, add 1/4 volume of 5× SDS Loading Buffer (0.25 % Bromophenol blue, 0.5 M DTT, 50 % Glycerol, 10 % SDS) to the "Input lysate" and "post-ChIP lysate" samples, add 1× SDS Loading Buffer to the "pre-wash beads" and "post-wash beads" samples to make the same volume as the "lysate" samples. Cover the "lysate" and "beads" samples with mineral oil and reverse crosslink the samples on a heat block set at 65 °C for at least 6 h. Load all samples side by side onto an 8 % SDS-PAGE gel with appropriate protein size markers. Perform Immunoblotting (Western blotting) to detect the amount of Piwi protein (detected as a band around 100 kDa) in each sample. A successful Piwi-ChIP will show a considerable amount of Piwi in the "pre-wash beads" and "post-wash beads" samples. In an ideal situation, the amount of Piwi protein in these "beads" samples should be comparable to that of the "Input lysate" sample, whereas the "post-ChIP lysate" sample should contain an undetectable amount of Piwi.
14. Always prepare and use fresh Elution Buffer.
15. Do not vortex.
16. For later application such as high-throughput sequencing, the DNA quantity of the final products can be measured by a regular quantitative PCR machine with Quant-iT Pico Green kit (from Life Technologies).

Acknowledgment

The development of the method described here was supported by NIH DP1CA174418 and an Ellison Medical Foundation Senior Scholar Award. We thank Ms. Ah Rume (julie) Park for valuable suggestions and Dr. Nils Neuenkirchen for proof-reading the manuscript.

References

1. Saxe JP, Lin H (2011) Small noncoding RNAs in the germline. Cold Spring Harb Perspect Biol 3:a002717
2. Pal-Bhadra M, Leibovitch BA, Gandhi SG, Rao M, Bhadra U, Birchler JA, Elgin SC (2004) Heterochromatic silencing and HP1 localization in Drosophila are dependent on the RNAi machinery. Science 303:669–672
3. Yin H, Lin H (2007) An epigenetic activation role of Piwi and a Piwi-associated piRNA in Drosophila melanogaster. Nature 450:304–308
4. Brower-Toland B, Findley SD, Jiang L, Liu L, Yin H, Dus M, Zhou P, Elgin SC, Lin H (2007) Drosophila PIWI associates with chromatin and interacts directly with HP1a. Genes Dev 21:2300–2311
5. Orlando V (2000) Mapping chromosomal proteins in vivo by formaldehyde-crosslinked-chromatin immunoprecipitation. Trends Biochem Sci 25:99–104
6. Megosh HB, Cox DN, Campbell C, Lin H (2006) The role of PIWI and the miRNA machinery in Drosophila germline determination. Curr Biol 16:1884–1894
7. Cox DN, Chao A, Lin H (2000) piwi encodes a nucleoplasmic factor whose activity modulates the number and division rate of germline stem cells. Development 127:503–514
8. Grimaud C, Bantignies F, Pal-Bhadra M, Ghana P, Bhadra U, Cavalli G (2006) RNAi components are required for nuclear clustering of Polycomb group response elements. Cell 124:957–971
9. Siomi MC, Saito K, Siomi H (2008) How selfish retrotransposons are silenced in Drosophila germline and somatic cells. FEBS Lett 582: 2473–2478
10. Schneider I (1972) Cell lines derived from late embryonic stages of Drosophila melanogaster. J Embryol Exp Morphol 27:353–365
11. Huang X, Yin H, Sweeney S, Raha D, Snyder M, Lin H (2013) A major epigenetic programming mechanism guided by piRNAs. Dev Cell 24:502–516
12. Shaffer CD, Wuller JM, Elgin SCR (1994) Preparation of Drosophila nuclei. Methods in Cell Biology 44:185–189

Chapter 2

Drosophila Germline Stem Cells for In Vitro Analyses of PIWI-Mediated RNAi

Yuzo Niki, Takuya Sato, Takafumi Yamaguchi, Ayaka Saisho, Hiroshi Uetake, and Hidenori Watanabe

Abstract

The *Drosophila piwi* gene has multiple functions in soma and germ cells. An in vitro system provides a powerful tool for elucidating PIWI function in each cell type using stable cell lines originating from germline stem cells (GSCs) and ovarian soma of adult ovaries. We have described methods for the maintenance and expansion of GSCs in an established cell line (fGS/OSS) and an in situ hybridization method for analyzing *piwi*.

Key words *Drosophila* germline stem cells, Ovarian cells, fGS/OSS

1 Introduction

Drosophila P-element induced wimpy testis (*piwi*) was first described as a gene required for the maintenance and normal development of germline stem cells (GSCs) in the testes and ovaries [1–3]. The *Drosophila* GSC niche is a suitable experimental model for RNAi study because it is accessible to various molecular and genetic tools [4]. Piwi is localized in the nuclei of somatic and germinal ovarian cells. The renewal of GSCs is maintained by *piwi* expression in the ovarian somatic niche [2, 3, 5]. An additional role of Piwi in germline development is the formation of maternally inherited polar plasma [6]. This gene also plays a role in Piwi-interacting RNA silencing, e.g., the suppression of transposable element expression [7–11].

Drosophila GSCs reside at the anterior tip of the germarium and are in direct contact with cap [12] and escort cells. One of the daughter cells of GSCs (called the cystoblast) leaves the niche and moves posteriorly with the help of escort cells [13, 14] (Fig. 1a). After four synchronous incomplete cycles of cytokinesis, the cystocyte (descendant from the cystoblast) is surrounded by

Mikiko C. Siomi (ed.), *PIWI-Interacting RNAs: Methods and Protocols*, Methods in Molecular Biology, vol. 1093, DOI 10.1007/978-1-62703-694-8_2, © Springer Science+Business Media, LLC 2014

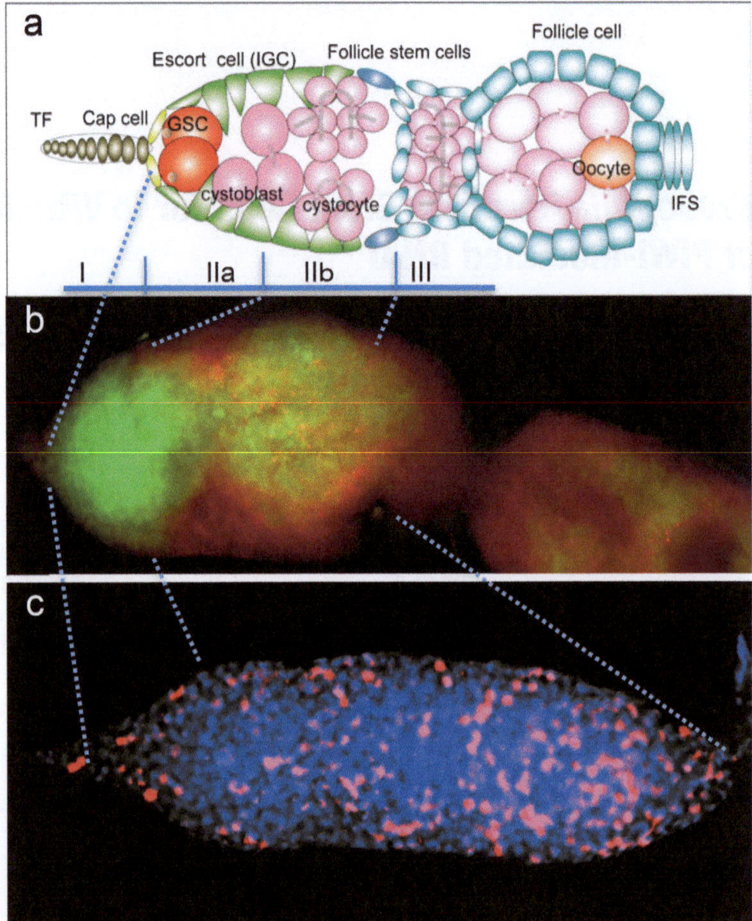

Fig. 1 Diagram of wild-type germarium and immunostained *bam* germaria. (**a**) Diagram of wild-type germarium. GSCs associating with cap and escort cells at the anterior tip undergo self-renewal division. One of the daughter cells (cystoblast) moves posteriorly, dissociates from the cap cells, and undergoes four cycles of synchronous, incomplete cytokinesis to produce 16-cell cystocytes. The 16-cell cystocytes are then surrounded with prefollicular cells derived from follicular stem cells. The follicular cells undergo eight mitotic divisions resulting in more than 600 cells. (**b**) Immunostained *bam* germarium with anti-Vasa (green) and Fas3 (red) antibodies, a germ cell and follicular cell marker, respectively. The hypertrophy of the germarium begins from just posterior to the follicular stem cell location where an egg chamber would be formed during normal oogenesis. (**c**) A mitotic *bam* germarium stained with anti-BrdU antibody (red) after BrdU incorporation. Note that mitotic cells are prominent at the posterior region of the germarium

prefollicular cells originating from two follicle stem cells [15] (Fig. 1a). The follicular cells undergo approximately eight mitotic divisions and produce more than 600 cells that surround each maturing egg chamber [15, 16].

Undifferentiated GSCs accumulate in the anterior and posterior regions of the germarium in ovarian tumors of the *bam-of-marble* (*bam*) mutant. These correspond to regions IIb and III of the normal germarium (Fig. 1b). In addition to GSCs, prefollicular cells continue to overproduce in the germarium as the female ages. Observation of the *bam* ovary stained with an anti-BrdU antibody after BrdU incorporation reveals prominent mitotic GSCs and soma located posteriorly (Fig. 1c). More than 1,000 cells, consisting of similar number of GSCs and prefollicular cells, accumulate in each ovariole of a 20-day-old female (unpublished). It is possible that the posterior expansion of GSCs is sustained by Dpp and other growth factors produced by prefollicular cells.

We successfully cultured adult GSCs after dissecting adult tumorous ovaries of the genotype of w^{1118}; *P[w hsp-70 bam*$^{+}$*]*$^{11\text{-}d}$ *bam*^{D86}*/bam*^{D86} *P[ovo-LacZ] P[vasa-egfp]* for a long term [17]. In addition, we established a cell line consisting of germ and soma cells and named it fGS/OSS. We then selected subpopulations consisting of only somatic cells by checking for the absence of vas-Egfp expression and established more than 20 independent subpopulations of somatic cells. There were no conspicuous variations of cellular morphology among these subpopulations. The somatic cells were flat and less than 10 μm in diameter. The uniformity of these subcultures indicates that the somatic cells were of one type and they originated from follicle stem cells and/or prefollicular cells. We named the somatic cell line OSS ([17, 18] and Fig. 2a, b). The OSS cells have a high mitotic activity that is similar to prefollicular cells in vivo. This cell line has been used for studying the biochemical nature of PIWI-related RNAi machinery in the ovarian soma [5, 19].

The maintenance and expansion of GSCs are sustained by growth factors supplied from OSS cells. GSCs and OSS cells directly associate with each other via DE-cadherin-mediated adherence junctions. In contrast to fGSCs, OSS cells can be maintained and expanded in culture under low nutrient conditions. Figure 2c shows an example of the growth curve of OSS cells with various concentrations of fly extract (FE) supplemented in the culture medium. Furthermore, the number of GSCs begins to decrease or is completely lost during rapid expansion of OSS cells. A possible explanation for this is the predominance of S- and M-phase OSS cells in the cell population. These cells would lose their functional ability to make contact with GSCs and produce sufficient quantities of growth factors. In addition GSC loss could be attributed to the formation of OSS cell clumps that surround GSCs as the former reach confluence (Fig. 2a, b). This would result in a lack of oxygen, nutrients, and space, all of which are necessary for survival and expansion of GSC. It is noteworthy that the contact inhibition observed in normal mammalian cell cultures does not occur in insect cultures. Thus, a key factor for the successful maintenance and expansion of GSCs in fGS/OSS culture is the suppression of their rapid expansion.

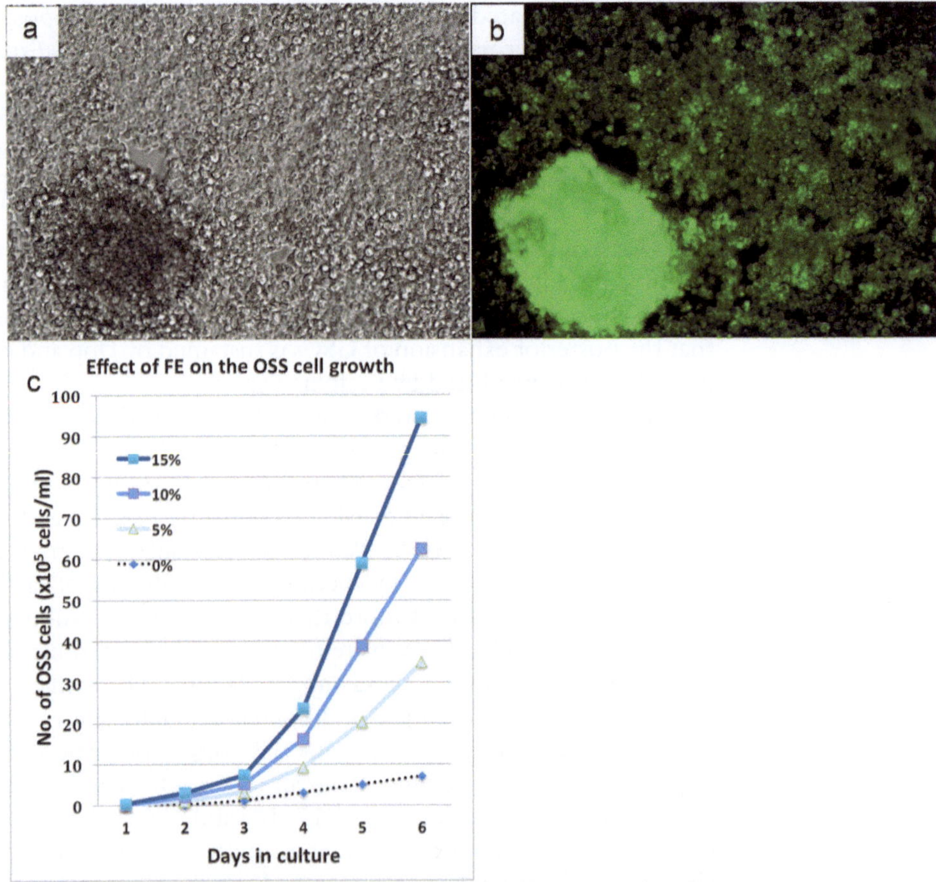

Fig. 2 Characteristics of fGS/OSS and OSS cells. Phase-contrast (**a**) and fluorescent (**b**) images of confluent fGS/OSS cells. As cells become confluent, a characteristic cell clump is formed in which GSCs are surrounded with OSS cells. (**c**) Effect of FE on the expansion of OSS cells. Since OSS cells originate from prefollicular cells with high mitotic activity, they expand rapidly even when maintained in low concentrations of FE. They expand more than 30 and 90 times in culture media supplied with 5 and 10 % FE, respectively, within 5 days

2 Materials

Prepare all solutions using Millipore-filtered ultrapure water. The solutions are freshly prepared before each use. Use nuclease-free water in each experiment and endotoxin-free water in cell culture.

2.1 Cell Culture

1. Shields and Sang M3 insect medium (Sigma).
2. L-Glutamic acid potassium salt monohydrate (Sigma).
3. Potassium bicarbonate (Sigma).
4. Glutathione* (Sigma).

5. Insulin** (Sigma).
6. Fetal bovine serum (FBS; endotoxin-free).
7. 1 % NaOH.
8. 1 N HCl.
9. Fly extract (FE; *see* Subheading 3.3).
10. 96-Well culture plate (Sumilon: Sumitomo Bakelite).
11. Syringe filter (33-mm Millex filter unit: Millipore).
12. 10-mL Luer-lok™ syringe.
13. 1,000-mL Stericup® filter unit (Millipore).
14. 150-mL Stericup® filter unit (SCGVU01RE, Millipore).
15. 15-mL conical tube.

**Stock solution of 100× glutathione*

1. Dissolve 6.0 g of glutathione in 100 mL Milli-Q water.
2. Filter sterilize and store as 1.0-mL aliquots at −20 °C.

***Stock solution of 100× insulin*

1. Dissolve 100 mg insulin in 100 mL Milli-Q water.
2. Add 100 μM HCl, filter sterilize, and store as 1.0-mL aliquots at −20 °C.

2.2 In Situ Hybridization

1. ddH_2O: Direct-Q®3 Ultrapure water system from Millipore.
2. *Drosophila* phosphate-buffered saline (PBS) (pH 7.4).
3. PBS plus Triton X-100 (PBT): 0.1 % Triton X-100 in 1× PBS.
4. Ethanol/PBT: 75 % ethanol in PBT, 50 % ethanol in PBT, and 25 % ethanol in PBT.
5. Fixative solution: 4 % (w/v) formaldehyde, 5 % (v/v) acetic acid, and 0.9 % (w/v) NaCl. Freshly prepared before use.
6. Hybridization buffer: 5× SSC, 50 % formamide, 100 μg/mL salmon sperm DNA, 50 μg/mL heparin, 0.1 % Triton X-100, and ddH_2O.
7. PBT/hybridization buffer: 25 % PBT in hybridization buffer, 50 % PBT in hybridization buffer, and 75 % PBT in hybridization buffer.
8. 5 % bovine serum albumin (BSA): 5 % (w/v) BSA in PBT. Freshly prepared before use.

2.3 Colorimetric Detection (Alkaline Phosphatase, Fig. 3a)

1. Anti-digoxigenin (DIG) antibody: Anti-DIG-AP Fab fragments from sheep (Roche).
2. NTMT: 100 mM NaCl, 100 mM Tris–HCl (pH 9.5), 50 mM $MgCl_2$, and 0.1 % Triton X-100. Store at 4 °C.

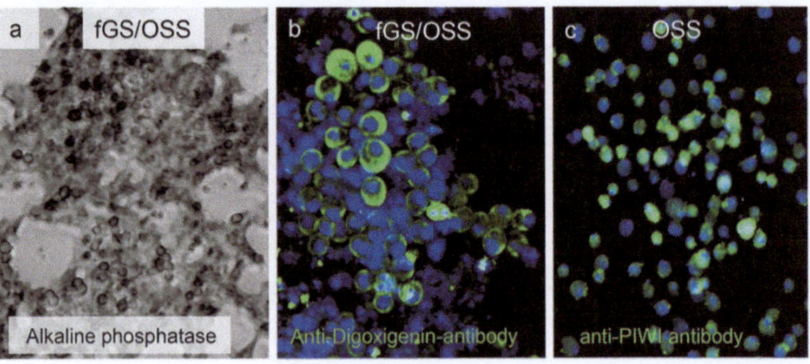

Fig. 3 *piwi* expression in fGS/OSS and OSS cells. Colorimetric detection of *piwi* mRNA with alkaline phosphatase (**a**) and fluorescence detection with anti-DIG–fluorescein antibody (**b**). (**c**) Immunostained OSS cells with anti-PIWI antibody. GSCs express higher levels of *piwi* mRNA than OSS cells. Weak signals in OSS cells stained with the anti-DIG-fluorescein antibody are masked by strong signals in GSCs. Accumulation of PIWI in the nuclei of OSS cells can be clearly observed when stained with the anti-PIWI antibody. Bar represents 10 μm

3. NTMT–nitroblue tetrazolium (NBT)/5-bromo-4-chloro-3-indolyl phosphate (BCIP): 4.5 μL/mL NBT (75 mg/mL NBT in 70 % dimethylformamide) and 3.5 μL/mL BCIP (50 mg/mL BCIP in dimethylformamide) in 1 mL NTMT. Solutions are freshly prepared before each use.

2.4 Fluorescence Detection (Fig. 3b)

1. Anti-DIG antibody: Anti-DIG-fluorescein Fab fragments from sheep (Roche). Store at 4 °C in the dark.

3 Methods

Perform all procedures at room temperature unless otherwise specified. Prepare and handle culture medium and cells at a clean bench or in a clean room.

3.1 Preparation of M3 (BF) Medium

Prepare the M3 (BF) medium according to the manufacturer's protocol.

1. Dissolve M3 powder in 700–800 mL sterilized Milli-Q water.
2. Add 1.0 g L-glutamic acid potassium salt monohydrate and 0.5 g potassium bicarbonate.
3. Adjust the medium to pH 6.8 with 1 % NaOH.
4. Add Milli-Q water to a final volume of 1 L.
5. Filter sterilize the medium through a 1,000-mL Stericup filter unit and store at 4 °C.

3.2 Preparation of Culture Medium for fGS/OSS and OSS Cells

1. Mix 50–60 mL of the above M3 (BF) medium, 10 mL of heat-inactivated FBS (pre-incubated at 50 °C for 10 min), and 1.0 mL each of 100× glutathione and 100× insulin. Make up to a final volume of 100 mL by the adding M3 (BF) medium.
2. Sterilize the mixed medium by passage through a 150-mL Stericup filter unit and store as 10-mL aliquots in 15-mL conical tubes at 4 °C.

3.3 Preparation of FE

FE is prepared according to the method of Currie et al. [20].

1. Prepare young adult flies of wild-type Oregon-R, aged 1–2 days post emergence. Two hundred flies are required for a 1.5-mL of aliquot.
2. Sterilize the flies with 70 % ethanol for 10 min.
3. Wash the sterilized flies three times with sterilized PBS and once with M3 (BF) medium supplemented with 10 % FBS.
4. Homogenize 200 sterilized flies with 1.2 mL of M3 (BF) medium containing 10 % FBS (1,000 flies with 6.0 mL of the culture medium) on ice.
5. Centrifuge the homogenate at 1,500 × *g* for 20 min at 4 °C.
6. Heat inactive the supernatant at 60 °C for 5–10 min.
7. Centrifuge the heat-inactivated supernatant at 6,000 × *g* for 10 min at 4 °C.
8. Transfer the supernatant into a fresh 1.5-mL tube and repeat centrifugation at 10,000 × *g* for 20 min.
9. Perform a second centrifugation of the supernatant at 15,000 × *g* for 60 min at 4 °C.
10. Make up to a final volume of 1.5 mL by adding M3 (BF) medium + 10 % FBS.
11. Store the FE aliquot at −20 °C after filter sterilization.

3.4 Maintenance and Expansion of fGS/OSS Cells

The most critical factor for the survival of GSCs is the supply of an adequate concentration of FE. A second limiting factor is cellular density as previously mentioned. Select GSC-rich subpopulations after splitting into smaller cell populations. Less than 100 cells are sufficient for expansion of fGS/OSS cells.

1. Filter sterilize the culture medium and FE mixture (*see* **Note 1**).
2. Inoculate the small cell population of fGS/OSS or OSS cells into one well of a 96-well culture plate.
3. Add the above mixed culture medium to a final volume of 200 μL in each well.
4. Maintain the plate in a Tupperware box or small container at 25 °C (*see* **Note 2**).

5. Check for cell viability and cell expansion. Within 24 h a healthy culture will contain somatic cells spreading from the cell masses and round GSCs in association with the soma.
6. Perform daily checks on the growth of GSCs on the somatic sheet.
7. Feed the cells every 5–7 days by replacing half of the culture volume with fresh culture medium.
8. After reaching confluence, the cells are transferred to a fresh well on the culture plate by gently scraping and pipetting the cellular sheet with the tip of a 200-μL micropipette (*see* **Note 3**).

3.5 In Situ Hybridization

Several modifications were applied to the procedures reported in previous studies [21–25] and to the manufacturer's protocol [26] in order to optimize the protocol for use in a 96-well culture plate. Addition or removal of reagents by pipette must be performed with care throughout the experiment to avoid detachment of adherent cells from the well. Note that Triton X-100 can result in the production of unwanted bubbles.

Cell Preparation

1. Inoculate a 96-well culture plate with a small population of cells (1×10^3 to 10^4 cells/mL) and culture for 24–48 h (*see* **Note 4**).
2. Remove the medium (*see* **Note 5**).
3. Fix the cells for 30 min in the fixative solution.
4. Perform a second fixation in 100 % ethanol for 10 min after removal of the first fixative solution.
5. Rehydrate the cells by passage through an ethanol/PBT gradient (75, 50, and 25 %) with 10 min in each solution.
6. Perform two 10-min washes in PBT.
7. Prehybridize by adding hybridization buffer for 10 min and then incubate for 60 min at 55 °C. This step is optional.
8. Remove the hybridization buffer.

Hybridization

9. Dilute the DIG-labeled RNA probe (*see* **Note 6**) in 100 μL of hybridization buffer (10–20 ng/well).
10. Denature the DIG-labeled RNA probe at 80 °C for 5 min and store the probe at 4 °C in a PCR machine until use.
11. Add 100 μL of the prepared probe to the cells and hybridize at 60 °C for 18 h.
12. Remove the probe.
13. Perform two 30-min washes in prewarmed hybridization buffer at 60 °C.

14. Rinse for 20 min in prewarmed 25 % PBT/hybridization buffer at 60 °C.
15. Rinse for 20 min in prewarmed 50 % PBT/hybridization buffer at 60 °C.
16. Rinse for 20 min in 75 % PBT/hybridization buffer.
17. Perform two 20-min rinses in PBT.
18. Block the cells in 5 % BSA for 30 min.
19. Remove the BSA solution.

Colorimetric Detection (Alkaline Phosphatase) (Fig. 3a)

20. Stain the cells for 60 min with a 1:500 dilution of anti-DIG-AP in BSA.
21. Perform four 20-min washes in PBT.
22. Perform three 5-min rinses in NTMT.
23. Add NTMT-NBT/BCIP in the dark. Check the cells every 5 min under the microscope.
24. Remove NTMT-NBT/BCIP after the desired incubation period (*see* **Note** 7).
25. Perform three 5-min washes in PBT (*see* **Note 8**).

Fluorescence Detection (Fig. 3b)

26. Stain the cells overnight with a 1:500 dilution of anti-DIG-fluorescein antibody in BSA at 4 °C.
27. Perform four 20-min washes in PBT.

4 Notes

1. The quality of FE may vary between laboratories. Thus, the optimal concentration of FE and FBS should be determined according to individual experimental design. For example, if the cells are intended for in situ hybridization and histochemistry, then the amount of FE and FBS required is less (<5 % each), whereas for biochemistry, 10–15 % FE and FBS would be required.
2. Add 200 μL sterilized Milli-Q water to the outermost wells of the plate to avoid desiccation.
3. Dissolve cell masses by treatment with 0.1 % trypsin + 0.02 % EDTA at 37 °C for 5 min when cell masses or cell clumps are tightly associated.
4. A confluence level of approximately one-third (4.0×10^5 cells/mL), allows easy capture of individual cell images at the end of the experiment.
5. Leave approximately 20 μL of the medium in the well. Removal of all the medium from the well is likely to lead to drying out and cell loss.

6. Prepare DIG-labeled RNA probes [26]. We use alkaline hydrolysis to produce 500-bp RNA probes.
7. Check under the microscope for NTMT-NBT/BCIP color changes in the samples.
8. Samples are washed at 4 °C in the dark because this helps the residual NTMT-NBT/BCIP to react with the probes more slowly.

Acknowledgment

This work was supported by Grant-in-Aid for Scientific Research (KAKENHI) on Innovative Areas, "Regulatory Mechanism of Gamete Stem Cells" (#24560219) to Y.N.

References

1. Lin H, Spradling AC (1997) A novel group of *pumilio* mutations affects the asymmetric division of germline stem cells in the *Drosophila* ovary. Development 124:2463–2476
2. Cox DN, Chao A, Baker J, Chang L, Qiao D, Lin H (1998) A novel class of evolutionarily conserved genes defined by *piwi* are essential for stem cell self-renewal. Genes Dev 12:3715–3727
3. Cox DN, Chao A, Lin H (2000) *piwi* encodes a nucleoplasmic factor whose activity modulates the number and division rate of germline stem cells. Development 127:503–514
4. Thomson T, Lin H (2009) The biogenesis and function of PIWI proteins and piRNAs: progress and prospect. Annu Rev Cell Dev Biol 25:355–376
5. Saito K, Inagaki S, Mituyama T, Kawamura Y, Ono Y, Sakota E, Kotani H, Asai K, Siomi H, Siomi MC (2009) A regulatory circuit for *piwi* by the large Maf gene *traffic jam* in *Drosophila*. Nature 461:1296–1299
6. Megosh HB, Cox DN, Campbell C, Lin H (2006) The role of PIWI and the miRNA machinery in *Drosophila* germline determination. Curr Biol 16:1884–1894
7. Sarot E, Payen-Groschêne G, Bucheton A, Pélisson A (2004) Evidence for a *piwi* dependent RNA silencing of the *gypsy* endogenous retrovirus by the *Drosophila melanogaster flamenco* gene. Genetics 166:1313–1321
8. Kalmykova AI, Klenov MS, Gvozdev VA (2005) Argonaute protein PIWI controls mobilization of retrotransposons in the *Drosophila* male germline. Nucleic Acids Res 33:2052–2059
9. Vagin VV, Sigova A, Li C, Seitz H, Gvozdev V, Zamore PD (2006) A distinct small RNA pathway silences selfish genetic elements in the germline. Science 313:320–324
10. Saito K, Nishida KM, Mori T, Kawamura Y, Miyoshi K, Nagami T, Siomi H, Siomi MC (2006) Specific association of Piwi with rasiRNAs derived from retrotransposon and heterochromatic regions in the *Drosophila* genome. Genes Dev 20:2214–2222
11. Brennecke J, Aravin AA, Stark A, Dus M, Kellis M, Sachidanandam R, Hannon GJ (2007) Discrete small RNA-generating loci as master regulators of transposon activity in *Drosophila*. Cell 128:1089–1103
12. Forbes AJ, Lin H, Ingham PW, Spradling AC (1996) *hedgehog* is required for the proliferation and specification of ovarian somatic cells prior to egg chamber formation in *Drosophila*. Development 122:1125–1135
13. Decotto E, Spradling AC (2005) The *Drosophila* ovarian and testis stem cell niches: similar somatic stem cells and signals. Dev Cell 9:501–510
14. Morris LX, Spradling AC (2011) Long-term live imaging provides new insight into stem cell regulation and germline-soma coordination in the *Drosophila* ovary. Development 138: 2207–2215
15. Margolis J, Spradling A (1995) Identification and behavior of epithelial stem cells in the *Drosophila* ovary. Development 121: 3797–3807
16. Calvi BR, Lilly MA, Spradling AC (1998) Cell cycle control of chorion gene amplification. Gene Dev 12:734–744
17. Niki Y, Yamaguchi T, Mahowald AP (2006) Establishment of stable cell lines of *Drosophila* germ-line stem cells. Proc Natl Acad Sci U S A 103:16325–16330

18. Niki Y (2009) Culturing ovarian somatic and germline stem cells of *Drosophila*. Curr Protoc Stem Cell Biol 10:2E.1.1–2E.1.9
19. Lau NC, Robine N, Martin R, Chung W-J, Niki Y, Berezikov E, Lai EC (2009) Abundant primary piRNAs, endo-siRNAs and microRNAs in a *Drosophila* ovary cell line. Genome Res 19:1776–1785
20. Currie DA, Milner MJ, Evans CW (1988) The growth and differentiation in vitro leg and wing imaginal disc cells from *Drosophila melanogaster*. Development 102:805–814
21. Raap AK, Van de Rijke FM, Dirks RW, Sol CJ, Boom R, Van der Ploeg M (1991) Bicolor fluorescence in situ hybridization to intron- and exon mRNA sequences. Exp Cell Res 197:319–322
22. O'Neill JW, Bier E (1994) Double-label in situ hybridization using biotin and digoxigenin tagged RNA probes. Biotechniques 17:870–875
23. Tautz D, Pfeiffle C (1989) A non-radioactive in situ hybridization method for the localization of specific RNAs in *Drosophila* embryos reveals translational control of the segmentation gene hunchback. Chromosoma 98:81–85
24. Spletter ML, Liu J, Liu J, Su H, Giniger E, Komiyama T, Quake S, Luo L (2007) Lola regulates *Drosophila* olfactory projection neuron identity and targeting specificity. Neural Dev 2:14
25. Groves A (1995) III. In situ hybridization on cultured cells. Anderson Lab in situ hybridization protocols. 7 Aug 2012. http://wmc.rodentia.com/docs/Big_In_Situ.html
26. Roche Applied Science (2008) V. RNA labeling by in vitro transcription of DNA with DIG, biotin, or fluorescein RNA labeling mix. DIG application manual for nonradioactive in situ hybridization, 4th edn. Roche Diagnostics GmbH, Germany, pp 49–52

Chapter 3

RNAi and Overexpression of Genes in Ovarian Somatic Cells

Kuniaki Saito

Abstract

Emerging evidence indicates that PIWI proteins, in collaboration with PIWI-interacting RNAs (piRNAs), play a critical role in retrotransposon silencing in *Drosophila* gonadal somatic and germ-line cells. The recent establishment of female germ-line stem cells/ovarian somatic sheet and its derivative cell line, ovarian somatic cells (OSCs), allows researchers to study the molecular functions of several protein factors involved in the primary piRNA pathway in *Drosophila*. Although transgene expression is difficult to achieve in gonad-derived cell lines, transfection of both expression vectors and knockdown reagents is highly effective in OSCs. Here, I focus on techniques that knockdown or overexpress genes of interest in OSCs.

Key words RNAi, Overexpression, OSC, Retrotransposon, piRNA, *Drosophila*, Cell line

1 Introduction

PIWI-interacting RNAs (piRNAs) are a group of small RNAs mainly expressed from retrotransposon-enriched genomic loci. piRNAs and their PIWI family protein partners have the ability to silence the expression of retrotransposon mRNAs and protect the germ-line genome [1–7]. Two main mechanisms of piRNA biogenesis have been proposed, termed primary and secondary piRNA biogenesis [8, 9]. In Drosophila, the primary processing pathway generates 23–29 nt piRNAs from long precursor piRNA transcripts in both germ-line and gonadal somatic cells [2, 10, 11]. Furthermore, piRNAs are amplified in a feed-forward loop, termed the secondary processing pathway, by Aubergine and Argonaute3, which are mainly expressed in germ-line cells [8, 9]. Genetic and biochemical approaches using Drosophila or mouse have revealed that dozens of factors are involved in these pathways. For example, Tudor domain-containing proteins, chaperon-related proteins, and mitochondria-localized proteins have been identified [2, 12–26]. While these approaches enable the identification of proteins that are essential for these pathways, the molecular functions of these

Mikiko C. Siomi (ed.), *PIWI-Interacting RNAs: Methods and Protocols*, Methods in Molecular Biology, vol. 1093,
DOI 10.1007/978-1-62703-694-8_3,

factors and the precise sequential order in which they act are not fully characterized.

The recent establishment of a coculture system consisting of female germ-line stem cells and ovarian somatic sheet cells (fGS/OSS) enable the molecular mechanisms operating in ovarian tissue to be analyzed [27]. In addition, an ovarian somatic cell line (OSC) has been established from the fGS/OSS parental cell line [11]. OSCs and OSSs share similar gene expression features that distinguish them from fGSs, indicating that OSCs and OSSs are derived from the somatic lineages of the ovary [11]. OSCs have been passaged for more than 4 years since establishment, and our culture medium is not suitable for growing fGSs; therefore, OSCs are highly likely to share distinct characteristics with OSSs.

Owing to the simplicity of homogenous cell culture, OSCs and OSSs have been utilized to investigate roles of the piRNA pathway [11, 22, 28–33]. In OSCs, Piwi is highly expressed, associated with primary piRNAs derived from flamenco or protein coding loci and is required for retrotransposon silencing [11]. Because neither aubergine nor argonaute3 are expressed in OSCs, they are utilized to study the molecular functions of factors involved in the primary piRNA pathway, in the knowledge that effects of the secondary piRNA pathway are excluded [11].

This chapter contains a preparation procedure for fly extract (FE) to facilitate and scale-up OSC culture. Also described are an efficient method for knockdown with siRNA and two protocols for OSC transfection with expression vectors.

2 Materials

2.1 Cell Culture

1. Ovarian somatic cells (OSCs).
2. Shields and Sang M3 insect medium (SIGMA).
3. M3/FCS: Shields and Sang M3 insect medium containing 10 % fetal calf serum (FCS).
4. Insulin (SIGMA).
5. L-Glutathione reduced (Sigma).
6. 0.05 % Trypsin–EDTA solution (Invitrogen).
7. 1× Phosphate-Buffered Saline (PBS).
8. 100-mm tissue culture dishes.
9. 60-mm tissue culture dishes.
10. 6-well multiwall tissue culture dishes.
11. 0.2-μm filters.
12. Metal mesh (AS ONE).

2.2 Nucleofection and Transfection

1. Nucleofector kit V (Lonza).
2. Xfect transfection reagent (Clontech).
3. Plasmid DNA: purified by anion-exchange chromatography (e.g., using Genopure Plasmid Purification Maxi Kit, Roche).
4. 50–100 pmol/μl siRNA duplex (Sigma or Invitrogen is recommended).
5. Nucleofector II device (Lonza).
6. Counting chamber (Invitrogen).

3 Methods

3.1 Culture of OSCs

The protocol for culturing OSCs is very similar to that described previously [27, 34]. Below is the procedure for the large-scale culture of OSCs.

3.1.1 Preparation of FE for OSCs

1. Collect Oregon R flies (*see* **Note 1**) and anesthetize them with diethylether.
2. Transfer the sleeping flies into a 50-ml tube and measure their weight (this should be approximately 3–8 g dry weight).
3. To clean the flies, add 40 ml 70 % ethanol and shake for 5 min on a shaker platform.
4. Transfer flies to a metal mesh and rinse several times with PBS to remove ethanol.
5. Absorb PBS by dabbing the bottom of metal mesh with hygroscopic paper.
6. Transfer flies to a mortar using a micropipette tip and add 10 ml of M3/FCS.
7. Homogenize flies vigorously (*see* **Note 2**) using a pestle with the mortar on ice.
8. Add M3/FCS to the homogenate to give a final concentration of 0.2 g fly weight per ml and transfer to a 50-ml centrifuge tube.
9. Spin at 6,000 × g for 10 min at 4 °C. Transfer the supernatant into a new 50-ml tube. Keep the supernatant at 0 °C by placing on ice.
10. Put 1-ml aliquots of supernatant in 1.5-ml tubes and heat at 60 °C for 5 min. Immediately after heat treatment, place on ice for 3 min (*see* **Note 3**).
11. Spin at 17,000 × g for 5 min at 4 °C. Combine the clear supernatants into a 50-ml centrifuge tube.

12. Centrifuge at 11,000 × *g* for 5 min at 4 °C. Collect the clear yellow colored supernatant (unsterilized fly extract, uFE) and store at −20 °C.

3.1.2 Preparation of Media Used for OSCs

1. Thaw frozen uFE in a 37 °C water bath. After thawing completely, place uFE on ice.
2. Sterilize uFE by passing through a 0.2-μm filter using aseptic technique, to yield sterilized fly extract (FE). In this step, several filters might be required. Approximately 10 ml of uFE solution can be filtered per filter.
3. Mix M3 media supplemented with 10 % FCS, 1× insulin, 1× glutathione, and 10 % FE in a plastic bottle.
4. Filter complete OSC media through a 0.2-μm filter (*see* **Note 4**). Store at 4 °C.

3.2 pAcM Vector to Overexpress Gene of Interest in OSCs

To express an exogenous gene, we constructed the pAcM vector in which a multiple cloning site is inserted between the Myc-tag sequence and the polyadenylation signal [11]. pAcM expression is driven by the actin 5c promoter, which is constitutively active in OSC and S2 cells (Fig. 1).

3.3 Nucleofection of OSCs with Plasmid Vector and siRNA

1. Cultivate the OSCs to be nucleofected in 90-mm dishes in complete medium.
2. Remove medium at late log phase, and rinse cells twice with 10 ml PBS.
3. Add 2 ml trypsin–EDTA solution to completely cover the cells. Discard trypsin–EDTA solution and incubate cells at 37 °C for 1 min (*see* **Note 5**).
4. Check that OSCs have released from the dish with an inverted microscope.
5. Add 5 ml medium to neutralize the trypsin and pipette up and down to make a single-cell suspension.
6. Count the cells with a counting chamber. Transfer 3×10^6 cells (*see* **Note 6**) into a 15-ml centrifuge tube and pellet the cells at 1,500 × *g* for 5 min at room temperature.
7. Remove supernatant carefully and add 100 μl Nucleofector solution V with 200 pmol siRNA duplex (for knockdown), 5 μg plasmid (for overexpression) (also *see* Subheading 3.2) or both (for rescue experiment). Pipette up and down to resuspend cells.
8. Transfer a 100-μl aliquot into a cuvette and place cuvette in the holder in the Nucleofector II device.
9. Pulse at program setting T-029 (Nucleofector II device) and return the nucleofected cells to a 6-well culture dish filled with 3 ml complete medium (*see* **Note 7**).

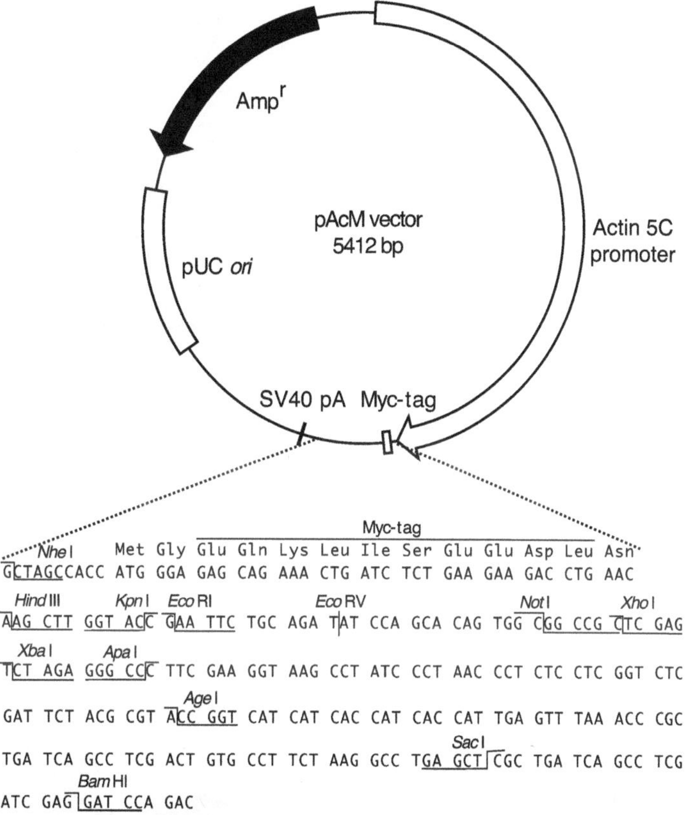

Fig. 1 pAcM vector for use in *Drosophila melanogaster*. pAcM is derived from pAc5.1-HisA1, a medium-copy ampicillin-resistance plasmid. The myc-tag was inserted and the multiple cloning site was modified. The actin 5C promoter is used to drive expression of a Myc-tag fusion protein. Unique restriction sites for insertion of sequences are shown. Using a maxi-prep protocol, 30–70 μg of plasmid can be obtained from a 250-ml culture

10. Incubate cells for 24–48 h at 26 °C and analyze RNA and/or protein levels, and consequences of gene overexpression (Fig. 2) or target gene depletion (Fig. 3).

3.4 Transfection of OSCs with Plasmid Vector

The Xfect transfection reagent was applied to introduce long dsRNA and plasmid into OSSs [30]. The protocol of transfection is very similar to that for nucleofection and is applicable for plasmid and long dsRNA transfection but not for siRNA transfection.

1. The day before transfection, trypsinize and count adherent cells in each well of a 6-well tissue culture dish. Seed 1×10^6 trypsinized cells in 2 ml of cell culture medium so that the cells will be 30–70 % confluent at the time of transfection.
2. Incubate at 26 °C overnight to allow cells to attach on the bottom of the dish.

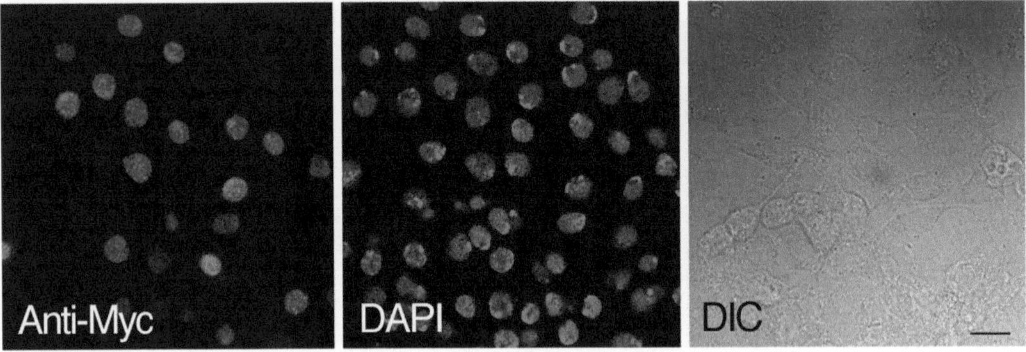

Fig. 2 Transient expression of Myc-tagged Piwi protein in OSCs. 48 h after nucleofection according to the protocol described here, cells were fixed and immunostained with anti-Myc antibody (*left*) and DNA was visualized with DAPI (*middle*)

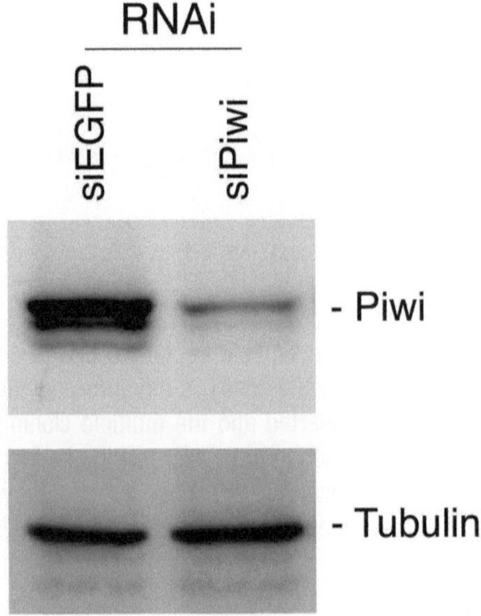

Fig. 3 Gene knockdown with siRNA duplex introduced by nucleofection. Piwi knockdown was performed using siRNA nucleofection. 48 h after nucleofection, Piwi expression was analyzed by western blotting using tubulin as an input control

3. On the day of transfection, add 4 μg plasmid DNA into 100 μl Xfect Reaction Buffer in a 1.5 ml tube (tube 1). In another 1.5 ml tube, mix 1.5 μl Xfect Polymer with 98.5 μl Xfect Reaction Buffer (tube 2).
4. Mix tube 1 with tube 2 and vortex for 10 s.
5. Incubate the sample for 10 min at room temperature.
6. During the incubation, wash OSCs with M3 media (containing neither FCS nor FE) and add 1 ml M3 media.

7. Add the 200 μl transfection solution and mix well.
8. Incubate the dish at 26 °C for 3 h (*see* **Note 8**).
9. Add 2.5 ml complete culture medium and incubate again at 26 °C.
10. Incubate 24–48 h at 26 °C and analyze RNA and/or protein levels, and consequences of target gene overexpression.

4 Notes

1. Oregon R flies are maintained in plastic vials by mass transfer of adults to fresh food. Adult flies kept at 25 °C are transferred into new vessels every 3–4 days (twice a week). 50–100 flies are transferred into each vial. We transfer 50–100 vials every transfer, so that we obtain 3–8 g dry weight of flies in a week. To make fly extracts, we use adult Oregon R flies at 4 days or at 7 days after eclosion.
2. It is important to emphasize that flies should be very well crushed. The extracts should be dark-red color.
3. Dividing the extract into 1-ml aliquots in 1.5-ml tubes is very important. When extracts were heat-treated the in a 50-ml tube, the resulting FE did not work well.
4. Our procedure does not adopt aseptic manipulation until 0.2-μm filtration. So, to reduce the risk of contamination, we filter the complete media before use.
5. Do not leave the cells to dry for more than 5 min, as this will lead to cells that are attached tightly to the dish.
6. When the experiment is scaled-up, 1×10^7 cells can be used in the same amount of siRNA and Nucleofector solution.
7. After nucleofection, you will see dead and damaged floating cells. The cell death rate depends upon the quality of the FE used in the OSC culture. Dead or floating cells should be less than 20 %.
8. OSCs are sensitive to loss of FE. If more than 10 % of cells are floating after this step, the incubation time should be decreased to 2 h or less.

Acknowledgments

This work was supported by the Japanese Society for the Promotion of Science (JSPS) through the Funding Program for Next Generation World-Leading Researchers (NEXT Program).

References

1. Ghildiyal M, Zamore PD (2009) Small silencing RNAs: an expanding universe. Nat Rev Genet 10:94–108
2. Malone CD, Hannon GJ (2009) Small RNAs as guardians of the genome. Cell 136: 656–668
3. Thomson T, Lin H (2009) The biogenesis and function of PIWI proteins and piRNAs: progress and prospect. Annu Rev Cell Dev Biol 25:355–376
4. Khurana JS, Theurkauf W (2010) piRNAs, transposon silencing, and Drosophila germline development. J Cell Biol 191:905–913
5. Saito K, Siomi MC (2010) Small RNA-mediated quiescence of transposable elements in animals. Dev Cell 19:687–697
6. Siomi MC, Sato K, Pezic D, Aravin AA (2011) PIWI-interacting small RNAs: the vanguard of genome defence. Nat Rev Mol Cell Biol 12:246–258
7. Pillai RS, Chuma S (2012) piRNAs and their involvement in male germline development in mice. Dev Growth Differ 54:78–92. doi:10.1111/j.1440-169X.2011.01320.x
8. Brennecke J, Aravin AA, Stark A, Dus M, Kellis M, Sachidanandam R, Hannon GJ (2007) Discrete small RNA-generating loci as master regulators of transposon activity in Drosophila. Cell 128:1089–1103
9. Gunawardane LS, Saito K, Nishida KM, Miyoshi K, Kawamura Y, Nagami T, Siomi H, Siomi MC (2007) A slicer-mediated mechanism for repeat-associated siRNA 5′ end formation in Drosophila. Science 315:1587–1590
10. Li C, Vagin VV, Lee S, Xu J, Ma S, Xi H, Seitz H, Horwich MD, Syrzycka M, Honda BM, Kittler EL, Zapp ML, Klattenhoff C, Schulz N, Theurkauf WE, Weng Z, Zamore PD (2009) Collapse of germline piRNAs in the absence of Argonaute3 reveals somatic piRNAs in flies. Cell 137:509–521
11. Saito K, Inagaki S, Mituyama T, Kawamura Y, Ono Y, Sakota E, Kotani H, Asai K, Siomi H, Siomi MC (2009) A regulatory circuit for piwi by the large Maf gene traffic jam in Drosophila. Nature 461:1296–1299
12. Lim AK, Kai T (2007) Unique germ-line organelle, nuage, functions to repress selfish genetic elements in *Drosophila melanogaster*. Proc Natl Acad Sci U S A 104:6714–6719
13. Pane A, Wehr K, Schupbach T (2007) zucchini and squash encode two putative nucleases required for rasiRNA production in the Drosophila germline. Dev Cell 12:851–862
14. Klattenhoff C, Xi H, Li C, Lee S, Xu J, Khurana JS, Zhang F, Schultz N, Koppetsch BS, Nowosielska A, Seitz H, Zamore PD, Weng Z, Theurkauf WE (2009) The Drosophila HP1 homolog Rhino is required for transposon silencing and piRNA production by dual-strand clusters. Cell 138:1137–1149
15. Nishida KM, Okada TN, Kawamura T, Mituyama T, Kawamura Y, Inagaki S, Huang H, Chen D, Kodama T, Siomi H, Siomi MC (2009) Functional involvement of Tudor and dPRMT5 in the piRNA processing pathway in Drosophila germlines. EMBO J 28: 3820–3831
16. Reuter M, Chuma S, Tanaka T, Franz T, Stark A, Pillai RS (2009) Loss of the Mili-interacting Tudor domain-containing protein-1 activates transposons and alters the Mili-associated small RNA profile. Nat Struct Mol Biol 16:639–646
17. Shoji M, Tanaka T, Hosokawa M, Reuter M, Stark A, Kato Y, Kondoh G, Okawa K, Chujo T, Suzuki T, Hata K, Martin SL, Noce T, Kuramochi-Miyagawa S, Nakano T, Sasaki H, Pillai RS, Nakatsuji N, Chuma S (2009) The TDRD9-MIWI2 complex is essential for piRNA-mediated retrotransposon silencing in the mouse male germline. Dev Cell 17: 775–787
18. Vagin VV, Wohlschlegel J, Qu J, Jonsson Z, Huang X, Chuma S, Girard A, Sachidanandam R, Hannon GJ, Aravin AA (2009) Proteomic analysis of murine Piwi proteins reveals a role for arginine methylation in specifying interaction with Tudor family members. Genes Dev 23:1749–1762
19. Frost RJ, Hamra FK, Richardson JA, Qi X, Bassel-Duby R, Olson EN (2010) MOV10L1 is necessary for protection of spermatocytes against retrotransposons by Piwi-interacting RNAs. Proc Natl Acad Sci U S A 107: 11847–11852
20. Patil VS, Kai T (2010) Repression of retroelements in Drosophila germline via piRNA pathway by the Tudor domain protein Tejas. Curr Biol 20:724–730
21. Zheng K, Xiol J, Reuter M, Eckardt S, Leu NA, McLaughlin KJ, Stark A, Sachidanandam R, Pillai RS, Wang PJ (2010) Mouse MOV10L1 associates with Piwi proteins and is an essential component of the Piwi-interacting RNA (piRNA) pathway. Proc Natl Acad Sci U S A 107:11841–11846
22. Handler D, Olivieri D, Novatchkova M, Gruber FS, Meixner K, Mechtler K, Stark A, Sachidanandam R, Brennecke J (2011) A systematic analysis of Drosophila TUDOR domain-containing proteins identifies Vreteno and the Tdrd12 family as essential primary piRNA pathway factors. EMBO J 30: 3977–3993
23. Qi H, Watanabe T, Ku HY, Liu N, Zhong M, Lin H (2011) The Yb body, a major site for

Piwi-associated RNA biogenesis and a gateway for Piwi expression and transport to the nucleus in somatic cells. J Biol Chem 286:3789–3797
24. Zamparini AL, Davis MY, Malone CD, Vieira E, Zavadil J, Sachidanandam R, Hannon GJ, Lehmann R (2011) Vreteno, a gonad-specific protein, is essential for germline development and primary piRNA biogenesis in Drosophila. Development 138:4039–4050
25. Anand A, Kai T (2012) The tudor domain protein kumo is required to assemble the nuage and to generate germline piRNAs in Drosophila. EMBO J 31:870–882
26. Preall JB, Czech B, Guzzardo PM, Muerdter F, Hannon GJ (2012) shutdown is a component of the Drosophila piRNA biogenesis machinery. RNA 18:1446–1457
27. Niki Y, Yamaguchi T, Mahowald AP (2006) Establishment of stable cell lines of Drosophila germ-line stem cells. Proc Natl Acad Sci U S A 103:16325–16330
28. Lau NC, Robine N, Martin R, Chung WJ, Niki Y, Berezikov E, Lai EC (2009) Abundant primary piRNAs, endo-siRNAs, and microRNAs in a Drosophila ovary cell line. Genome Res 19:1776–1785
29. Robine N, Lau NC, Balla S, Jin Z, Okamura K, Kuramochi-Miyagawa S, Blower MD, Lai EC (2009) A broadly conserved pathway generates 3′UTR-directed primary piRNAs. Curr Biol 19:2066–2076
30. Haase AD, Fenoglio S, Muerdter F, Guzzardo PM, Czech B, Pappin DJ, Chen C, Gordon A, Hannon GJ (2010) Probing the initiation and effector phases of the somatic piRNA pathway in Drosophila. Genes Dev 24:2499–2504
31. Olivieri D, Sykora MM, Sachidanandam R, Mechtler K, Brennecke J (2010) An in vivo RNAi assay identifies major genetic and cellular requirements for primary piRNA biogenesis in Drosophila. EMBO J 29:3301–3317
32. Saito K, Ishizu H, Komai M, Kotani H, Kawamura Y, Nishida KM, Siomi H, Siomi MC (2010) Roles for the Yb body components Armitage and Yb in primary piRNA biogenesis in Drosophila. Genes Dev 24: 2493–2498
33. Olivieri D, Senti KA, Subramanian S, Sachidanandam R, Brennecke J (2012) The cochaperone shutdown defines a group of biogenesis factors essential for All piRNA populations in drosophila. Mol Cell 47:954–969
34. Niki Y (2009) Culturing ovarian somatic and germline stem cells of Drosophila. Curr Protoc Stem Cell Biol. Chapter 2, Unit 2E 1

Chapter 4

Making piRNAs In Vitro

Shinpei Kawaoka, Susumu Katsuma, and Yukihide Tomari

Abstract

Classical biochemical approaches have made great contribution to our current understanding of how small interfering RNA (siRNA) and microRNA (miRNA) are produced and function. However, it has been challenging to establish in vitro systems that can dissect the mechanism of PIWI-interacting RNAs (piRNAs), largely due to the lack of suitable cultured cell lines possessing the piRNA pathway endogenously. The silkworm ovary-derived BmN4 cell line is an emerging model for studying piRNAs, because this cell line harbors a fully functional germline piRNA pathway. We recently reported in vitro recapitulation of a part of piRNA biogenesis—loading of precursor piRNAs and 3′-end maturation—by using lysate prepared from BmN4 cells. Here we describe the methods for maintenance of BmN4 cells, preparation of cell lysate, and in vitro assays to dissect piRNA biogenesis.

Key words PIWI, PIWI-interacting RNAs, Silkworm, BmN4 cell line, Cell culture, In vitro system

1 Introduction

PIWI-interacting RNAs (piRNAs) are small RNAs that bind to PIWI subfamily proteins, a subclade of Argonaute family proteins [1–4]. The piRNA pathway components generally show a gonad-restricted expression. Consistent with this, mutations in the piRNA pathway genes cause developmental defects in germ line cells, often resulting in sterility.

Biogenesis of piRNAs does not require the activity of double-strand RNA-specific Dicer proteins, and thus it is thought that piRNAs are born single-stranded [5, 6]. piRNAs are generated from the piRNA-producing loci dispersed in the genome called piRNA clusters [1, 3, 4, 7]. The single-stranded piRNA primary transcripts are likely fragmented into shorter single-stranded RNAs (designated as precursor piRNAs or pre-piRNAs) that are still longer than the mature piRNA length. Among pre-piRNAs, those beginning with uridine at their 5′ ends (1U) are selectively incorporated into a PIWI protein, Siwi in the silkworm [8–10]. Interaction between the 5′-end of pre-piRNAs and Siwi requires

Mikiko C. Siomi (ed.), *PIWI-Interacting RNAs: Methods and Protocols*, Methods in Molecular Biology, vol. 1093, DOI 10.1007/978-1-62703-694-8_4,

the phosphate-binding pocket in the MID domain of Siwi [9]. After pre-piRNA incorporation into Siwi, the extending 3′ tail is trimmed by an as-yet-unidentified 3′–5′ exonuclease, called Trimmer, into the mature piRNA length. Coupled with the trimming reaction, a methyltransferase Hen1 catalyzes 2′-*O*-methylation at the 3′-end of the trimming products [5–7, 9, 11–15], completing the formation of 1U primary piRNA-Siwi effector complex.

In insects, 1U primary piRNAs are often antisense to transposons, thereby directing associated PIWI proteins to transposon mRNAs. PIWI proteins, with the aid of their endonuclease activity, slice transposon mRNAs to silence them [16–20]. Interestingly, 3′-fragments of the sliced products are not simply degraded, but instead are loaded into another PIWI protein, BmAgo3 in the silkworm [8–10]. These RNA fragments likely serve as new pre-piRNAs and are further trimmed into the mature length and methylated at the 3′ end to produce secondary piRNAs. Therefore, in the piRNA pathway, the transposon-silencing step is tightly coupled with production of secondary piRNAs. Slicer-dependent piRNA biogenesis is generally called the ping-pong cycle, ping-pong amplification loop, or simply secondary piRNA biogenesis.

The above-described model for 3′-end maturation of piRNAs was established based on in vitro experiments using a silkworm ovary-derived BmN4 cell line, which harbors a fully functional germline piRNA pathway [8, 9]. Here we describe the methods for maintenance of BmN4 cells, preparation of lysates, and in vitro assays to detect 3′ end trimming and methylation.

2 Materials

2.1 Preparation of Cell Culture and Transfection

1. IPL-41 medium (Applichem): Thoroughly dissolve the powder in 800 mL of sterile water according to the instruction for more than 30 min at room temperature, adjust the pH to 6.2–6.4 by KOH or NaOH, add 0.35 g of $NaHCO_3$, 2.8 g of NaCl, and finally add sterile water to 1 L.
2. Fetal bovine serum (Gibco).
3. 1,000 mL Vacuum Filter/Storage Bottle System, 0.22 μm Pore 54.5 cm^2 CA Membrane, Sterile (Corning).
4. 75 cm^2 Rectangular Canted Neck Cell Culture Flask with Plug Seal Cap (Corning).
5. 25 cm^2 Rectangular Canted Neck Cell Culture Flask with Plug Seal Cap (Corning).
6. Six-Well Clear TC-Treated Multiple Well Plates, Individually Wrapped, Sterile (Corning).
7. 15 mL clear polypropylene centrifuge tubes (Corning).
8. Cellfectin®II reagent (Invitrogen).

9. Zeocin (Invitrogen).
10. pIZ-Flag/His-Siwi plasmid described in ref. [8].
11. Solution A: 2 μg of the plasmid and 100 μl of serum-free IPL-41.
12. Solution B: 10 μl of Cellfectin®II and 100 μl of serum-free IPL-41.
13. Polystyrene tube (Falcon).
14. Fetal Bovine Serum (Gibco).

2.2 Preparation of Lysate

1. 1× lysis buffer: 30 mM HEPES (pH 7.4), 100 mM KOAc, 2 mM $Mg(OAc)_2$. Store at 4 °C.
2. 1 M DTT: Dissolve in water and store in aliquots at −20 °C.
3. 25× protease inhibitors cocktail (25× PIC): Dissolve one tablet of Complete EDTA-free (Roche) in 2 mL water and store in aliquots at −20 °C.
4. Dounce homogenizer (WHEATON; 7 mL, "TIGHT" pestle).

2.3 Preparation of In Vitro Loading and Trimming Assays

1. Synthetic single-stranded RNAs listed in Table 1.
2. [γ-^{32}P] ATP (MP Biomedicals).
3. 100 mM cold ATP: Dissolve in water, adjust pH to 7–8 with KOH, and store in aliquots at −20 °C.
4. 20 mg/ml Glycogen (Roche): Store in aliquots at −20 °C.
5. T4 Polynucleotide kinase (PNK) (Takara).
6. G25 MicroSpin column (GE).
7. 40× reaction mix: 120 mL of 40× reaction mix contains 50 mL of water, 20 μL of 500 mM creatine monophosphate (Fluka; prepared fresh from powder), 20 μL of 1 mM amino acid stock (Sigma; 1 mM each amino acid), 2 μL of 1 M DTT, 1 μL of 40U/μL RNasin Plus (Promega), 4 μL of 100 mM ATP, 1 μL of 100 mM GTP, 16 μL of 1M KOAc, 6 μL of 2U/μL creatine phosphokinase (Cal-biochem; freshly prepared by diluting 2 μL of a 10 U/μL stock in 8 μL of 1× lysis buffer).

Table 1
Synthetic RNAs used in this study

Name	Seq (5′–3′)
1U-26	UCAAAAACUAACGGAUUGGUUUCGAA
1A-26	ACAAAAACUAACGGAUUGGUUUCGAA
1G-26	GCAAAAACUAACGGAUUGGUUUCGAA
1C-26	CCAAAAACUAACGGAUUGGUUUCGAA
1U-50	UCAAAAACUAACGGAUUGGUUUCGAACAGUCACCCGCCCGGACAGGUCCC

8. Dynabeads (Invitrogen).
9. Magnetic stand (Invitrogen).
10. Anti-Flag antibody (Sigma).
11. 0.5× TBE: 45 mM Tris, 45 mM Boric acid, 1 mM EDTA (pH 8.0). Store at room temperature or 4 °C.
12. Urea-containing Dye: 98 % (w/v) Formamide, 10 mM EDTA, 0.025 % (w/v) XC, 0.025 % (w/v) BPB.
13. 20 mg/ml Proteinase K (Nacalai): dissolve in 1× lysis buffer, 50 % glycerol.
14. 2× proteinase K buffer: 200 mM Tris pH 7.5, 25 mM EDTA, 340 mM NaCl.

2.4 Preparation of β-Elimination Followed by Oxidation

1. 5× borate buffer: 148 mM Borax, 148 mM Boric acid (pH 8.6).
2. 200 mM $NaIO_4$: Use only freshly dissolved $NaIO_4$.
3. Glycerol.
4. 500 mM NaCl.
5. 500 mM NaOH.
6. 20 mg/ml Glycogen (Roche): Store in aliquots at −20 °C.
7. *S*-adenosyl homocysteine (Sigma).

3 Methods

3.1 Generation of BmN4 Cells Stably Expressing Flag-Tagged Siwi

BmN4 cells are an adhesive type of cells, which are usually cultured in 75 cm² flasks typically with 8 mL of IPL-41 containing 10 % FBS prepared by following the manufacture's instruction. Once BmN4 cells form a single layer over the entire surface area available for cell growth, they start to form a "second layer" or "cell clusters," which can be easily collected by a gentle pipetting (Fig. 1, *see* **Note 1**). To split cells into a new flask or dish, discard old media, add 8 mL of fresh media, and gently pipette a surface area (defined as 1× cell mix). Generally, when splitting 1/2× cells into a new 75 cm² flask, it will take 3–4 days for cells to form a single layer, and another 3–4 days to form a second layer to be considered "ready-to-use" (*see* **Note 2**). For maintenance, a second layer should be washed by pipetting every 3–4 days to prevent over growth of cells.

1. To make a stably transfected cell line, split cells with a density of 1/2–1/4 dilution (generally ~2.5×10^5 cells/mL) to a 25 cm² flask with 4 mL of fresh media (*see* **Note 3**). Wait for a day to let cells be attached firmly to the bottom of the flask.
2. On the next day, remove media and add 1.8 mL of serum-free IPL-41 media. Prepare two tubes to make solution A and solution B of Cellfectin II reagent in the polystyrene tubes.

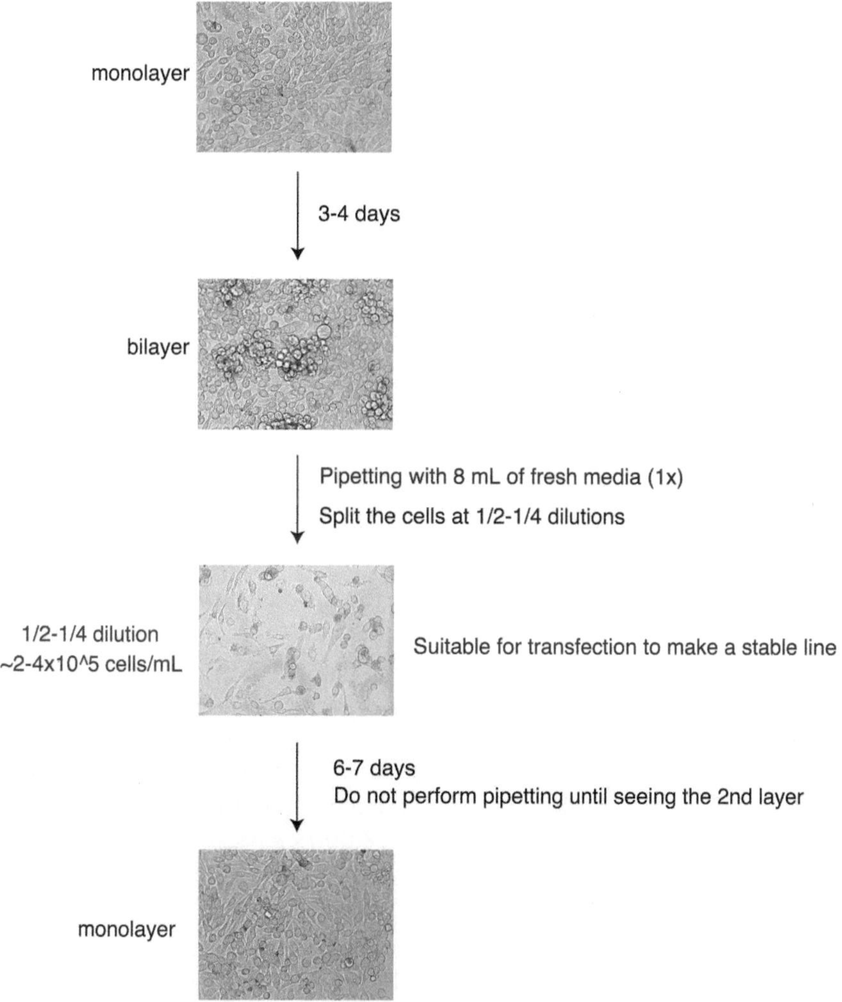

Fig. 1 Standard protocol for BmN4 cell culture. The pictures show representative conditions of BmN4 cells. See the text for details

Make sure to mix the reagents well. Mix solution A and solution B and wait for 20 min at room temperature. Gently add the mixture to the flask and incubate for 8 h.

3. Then, remove the media containing transfection complexes and add 4 mL of fresh, FBS-containing IPL-41. To generate cells stably expressing the zeocin-resistant gene cassette, culture cells in the same media for 2–3 days and then add zeocin with the final concentration of 500 μg/mL.

4. Change media every 3–4 days. When the transfection is successful, clusters of growing cells could be observed even in the presence of zeocin. It generally takes 2–3 weeks to obtain the first layer of BmN4 cells resistant to zeocin.

5. Expand cells into a 75 cm^2 flask, and check the expression of transgene by Western blot or qRT-PCR analysis. After the selection, the final concentration of zeocin can be reduced to 250 μg/mL. We recommend confirming the expression of transgene at least every 2 months, because sometimes cells somehow lose the transgene expression, even though they still retain the zeocin resistance.

3.2 Preparation of Lysate for In Vitro Loading and Trimming Assays

1. Collect cells stably expressing Flag-Siwi proteins into a 50 mL conical tube, spin down cells (2,000 × *g*, 20 °C, 10 min), and resuspend in ~1 mL of ice-cold PBS.
2. Measure the weight of an empty 1.5 mL tube (record as *X*) prior to transferring cells into it. Transfer cells into the 1.5 mL tube, and perform centrifugation (2,000 × *g*, 20 °C, 5 min). Remove supernatant, and measure the weight of the pellet + tube (record as *Y*).
3. Add the equal volume to the cells (($Y-X$) μl) of 1× lysis buffer supplemented with the proteinase inhibitor cocktail and DTT (f.c. 1 mM), and resuspend them by gentle pipetting.
4. Transfer cells into the Dounce homogenizer on ice or in a cold room. Carefully stroke the homogenizer for 20 times, and transfer homogenate into the 1.5 mL tube. Perform centrifugation (1,000 × *g*, 4 °C, 20 min). Collect supernatant as 1,000 × *g* lysate, and centrifuge these again (17,000 × *g*, 4 °C, 20 min) to obtain 17,000 × *g* lysate (supposed to be cytoplasmic fraction) containing the majority of Flag-Siwi proteins (*see* **Note 4**).
5. Resuspend the pellets with the same volume of 1× lysis buffer supplemented with proteinase inhibitor cocktail and DTT (1,000 × *g* pellet). To avoid freeze and thaw many times, it is recommended to transfer 30 μl aliquots of lysate into 0.2 mL tubes, freeze them by liquid nitrogen, and store at −80 °C until use. Because expression and activity of Flag-Siwi proteins may vary in different preparation batches, it is recommended to make a large stock of lysate at once. We previously showed that Flag-Siwi proteins active for pre-piRNA loading are enriched in 17,000 × *g* lysate, while 3′-end trimming activity resides in 1,000 × *g* pellet [9]. Thus, 17,000 × *g* lysate is used for the loading assay and 1,000 × *g* pellet is used for the trimming assay in the following protocol.

3.3 Preparation of Radiolabeled RNA for In Vitro Loading and Trimming Assays

Synthetic single-stranded (ss) RNAs used in this study are listed in Table 1.

1. To label the 5′ end of the ssRNAs with ^{32}P, mix 1 mL of 10 mM single-stranded guide strand RNA, 0.7 μL of [γ-^{32}P]ATP, 1 μl of T4 PNK, 2 μl of PNK reaction buffer, and 15.3 μl of water.

Mix the reaction well and incubate the reaction at 37 °C for more than 1 h.

2. Then, to remove the unincorporated [γ-^{32}P]ATP, adjust the volume to 40 μl with water and spin through a G-25 MicroSpin column.
3. Precipitate the column flow-through with 1/10× volume of 3 M NaOAc (pH 5.5), 1 μl of 20 mg/ml glycogen and 3× volume of absolute ethanol, and rinse the pellet with 70 % ethanol.
4. Finally, dissolve the precipitate in 10 μl of water, which will make the final concentration of 1 μM.
5. Take 0.2 μl of radiolabeled ssRNAs into 100 μl of urea-containing dye, and run 5 μl on 8 M urea-containing 12 % acrylamide gel. Dry the gel by using a gel dryer at 90 °C, and perform autoradiography for checking the purity and quantity of radiolabelled ssRNAs. This is especially important when comparing the loading efficiency among different ssRNAs (e.g., those starting with 5′ U/A/G/C). Although radioactivities of different ssRNAs should usually be very close, some variations might be observed. In such a case, dilute the samples to normalize the radioactivities. The labeled ssRNAs can be stored at −20 °C until use.

3.4 In Vitro Loading Assay

It is critical to define the "canonical loading" of piRNAs in vitro. Most piRNAs begin with U at the 5′ ends. One objective criteria we can use for defining canonical loading is the "1U-bias" of piRNAs.

1. Thaw 26-mer ssRNAs with 1U/1A/1G/1C and 17,000×g lysate on ice. Empirically, Siwi proteins without ssRNAs are very unstable at room temperature, e.g., incubation of naked Siwi proteins at room temperature without exogenous ssRNAs for more than 30 min could result in loss of the 1U-bias in this assay.
2. Prepare 10 μl of Dynabeads per one reaction, equilibrate the beads by washing with 1× lysis buffer supplemented with DTT (f.c. 1 mM) twice. Resuspend the beads with 10 μl of 1× lysis buffer supplemented with DTT, and add 1 μl of anti-Flag antibody to the beads. After incubating the beads with the antibody for more than 30 min on ice, place the tube in the magnetic stand, and remove supernatant containing the free antibody.
3. Then, resuspend the beads with ~20 μl of 1× lysis buffer supplemented with DTT. Once the beads with anti-FLAG antibody are prepared, mix 3 μl of 40× reaction mix, 6 μl of 17,000×g lysate, and 1 μl of ssRNA per one reaction.

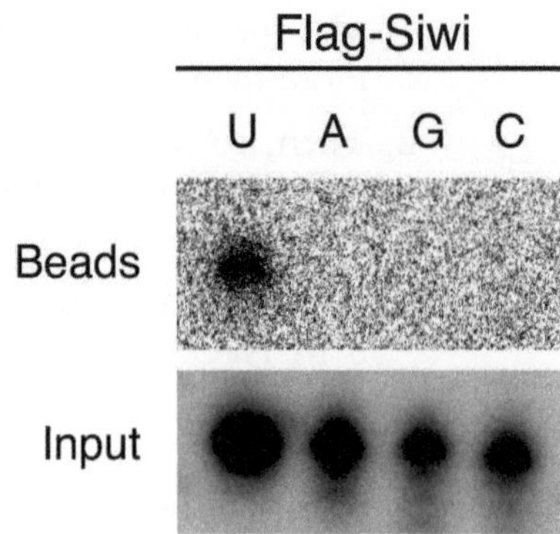

Fig. 2 In vitro loading assay. 17,000 × *g* lysate from cells expressing Flag-Siwi was incubated with radiolabelled 26 mer synthetic RNAs beginning with U, A, G, or C. The reactions were immunoprecipitated with anti-Flag antibody

4. Incubate the reaction mix for 30 min at 25 °C. Then, add the beads to the reaction mix, and incubate for more than 1 h on ice with occasional tapping.
5. Remove the supernatant and wash the beads with 1× lysis buffer supplemented with DTT and 0.1 % Empigen three times. Add 100 μl of the proteinase K reaction mix containing 100 μl of 2× proteinase K buffer, 10 μl of 20 mg/ml of proteinase K, and 1 μl of 20 mg/ml glycogen, and incubate for 15 min at 65 °C.
6. Place the tube on the magnetic stand, and transfer supernatant into the new 1.5 mL tube. Add 3× volume of absolute ethanol and perform centrifugation (17,000×*g*, 4 °C, 15 min). Remove supernatant, dry up, and dissolve the pellet with 10 μl of 1× loading dye. Load 4.5 μl of samples onto 8 M Urea-containing 12 % denature acrylamide gel (Fig. 2).

3.5 In Vitro Trimming Assay

Trimming assay is also a good indicator for canonical loading, because longer ssRNAs canonically loaded into Siwi proteins should be appropriately trimmed. As mentioned above, the loading reaction is sensitive to the lysate condition and requires careful handing on ice. On the other hand, trimming lysate (1,000×*g* pellet) seems quite stable even at room temperature. The protocol for making trimming substrate—Flag-Siwi/1U-50 RNA complexes—is essentially the same as that for the in vitro loading assay.

1. Briefly, incubate 17,000×*g* lysate containing Flag-Siwi with radiolabelled 50-mer 1U RNAs in the presence of 40× reaction mix, immunoprecipitate the complexes with anti-Flag antibody immobilized on magnetic beads.
2. After washing the beads with 1× lysis buffer supplemented with DTT and 0.1 % Empigen three times, wash the beads

with 1× lysis buffer supplemented with DTT to wash out the remaining Empigen.

3. Then, add 7 μl of 1,000×*g* pellet (or 1× lysis buffer as negative controls) and 3 μl of 40× reaction mix, and incubate the reaction mix at 25 °C (*see* **Note 5**).
4. Add 100 μl of the proteinase K reaction mix containing 100 μl of 2× proteinase K buffer, 10 μl of 20 mg/ml of proteinase K, and 1 μl of 20 mg/ml glycogen to the beads, and incubate for 15 min at 65 °C.
5. Place the tube on the magnetic stand, and transfer supernatant into the new 1.5 mL tube. Add 3× volume of absolute ethanol and perform centrifugation (17,000×*g*, 4 °C, 15 min). Remove supernatant, dry up, and dissolve the pellet with 10 μl of 1× loading dye. Load 5 μl each of the samples onto 8 M Urea-containing 12 % denature acrylamide gel (Fig. 3).

3.6 Analysis of the 3′ End of Trimmed RNAs

To conclude the trimming products are canonical, mature piRNAs, it is not enough to see the length of trimming products. One way to confirm that the products are canonical piRNAs is to investigate the 3′ end modification of the trimming products. To test this, β-elimination followed by $NaIO_4$ oxidation can be utilized. Unmodified RNAs are sensitive to the reaction, being converted into one-nucleoside shorter RNA with 3′-monophosphate

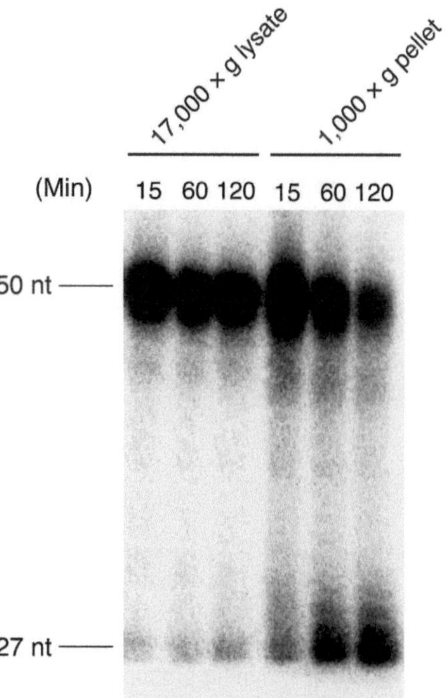

Fig. 3 Trimming assay. Immunopurified Flag-Siwi/1U-50 complexes were incubated with 17,000 × *g* lysate or 1,000 × *g* pellet

by the reaction. The resulting RNAs migrate faster in polyacrylamide gel electrophoresis. In contrast, 3′-modified RNAs are resistant to the reaction.

1. Dissolve the pellet of the trimmed RNAs with 10 μl of water instead of loading dye. Use one half for $NaIO_4$ reaction and another half for a negative control.
2. Add 12.5 μl of 200 mM $NaIO_4$, 40 μl of 5× borate buffer to the resultant RNA, and adjust the reaction to in total 100 μl with water. For a negative control, add 20 μl of water instead of $NaIO_4$ solution.
3. Incubate the reaction for 30 min at room temperature. Add 10 μl of glycerol to the reaction to quench unreacted $NaIO_4$, and then induce β-elimination by adding 10 μl of 500 mM NaOH. Incubate the samples at 45 °C for 90 min.
4. To precipitate the RNA, add 180 μl of 500 mM NaCl, 1 μl of glycogen, and 900 μl of the absolute EtOH, followed by centrifugation (17,000 × *g*, 4 °C, 15 min).
5. Remove the supernatant, wash the pellet with 500 μl of 70 % EtOH, and perform centrifugation (17,000 × *g*, 4 °C, 5 min). Dissolve the resultant pellet with 10 μl of loading dye, and run 5 μl of samples on 8M Urea-containing 12 % denature acrylamide gel (Fig. 4, *see* **Note 6**). As shown in Fig. 4a, untrimmed 50-nt RNAs are sensitive to the reaction and migrated faster while the trimming products are resistant, suggesting that 3′-end of trimming products are blocked. *S*-adenosyl homocysteine (SAH) is a competitive compound for *S*-adenosyl methionine (SAM) that is required for 2′-*O*-methylation.

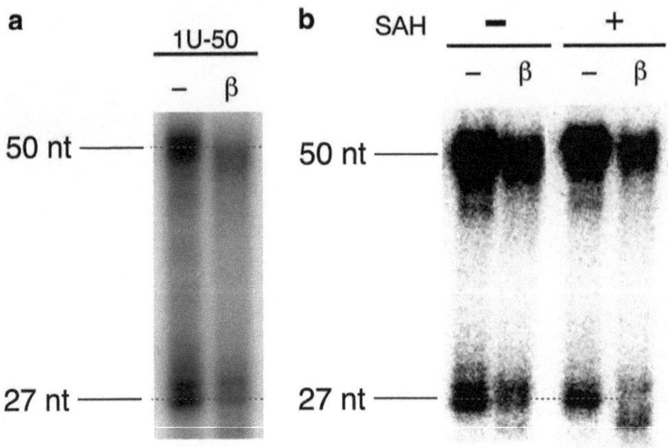

Fig. 4 Testing 3′-end modification of trimming products. (**a**) 3′ modification state of trimming products was analyzed by β-elimination followed by oxidation. (**b**) SAH, an inhibitor for 2′-*O*-methylation reaction at the 3′ end of piRNAs, was added in trimming reaction. 3′ modification state of trimming products was analyzed by β-elimination followed by oxidation

Addition of SAH in the trimming reaction (f.c. 1 mM) clearly inhibited 3′-end blockage for trimming products (Fig. 4b). Thus, it is very likely that 3′-end blockage of trimming products are mediated by 2′-*O*-methylation.

4 Notes

1. Generally we perform pipetting 20 times with a weak power setting. It is recommended to confirm the removal of the second layer under the microscope. When using the same flask for more than a month, it may become difficult to peal off the second layer by pipetting (*see* **Note 2**).
2. Related to **Note 1**, we recommend transferring cells to a new flask every 1 month.
3. For drug selection, cells should be transfected at a lower density compared to transient transfection. Cells seeded at a higher density might somehow propagate without having a drug resistant gene cassette.
4. If 1,000× lysate (required for the trimming assay) is not necessary, proceed to 17,000 ×*g* centrifugation without performing 1,000 ×*g* centrifugation.
5. Trimming activity of 1,000 ×*g* pellet can vary depending on the lysate batch. Generally, 1 h reaction results in at least 50 % trimming in the assay.
6. The pellets after $NaIO_4$/β-elimination usually become thin and broad. Thus dissolve them carefully not to miss the samples stuck to the tubes.

Acknowledgments

We thank Keisuke Shoji, Maki Kobayashi, Mayu Yoshikawa, and the members of our laboratory for the critical comments and helps on the manuscript.

References

1. Malone CD, Hannon GJ (2009) Molecular evolution of piRNA and transposon control pathways in *Drosophila*. Cold Spring Harb Symp Quant Biol 74:225–234
2. Slotkin RK, Martienssen R (2007) Transposable elements and the epigenetic regulation of the genome. Nat Rev Genet 8:272–285
3. Ghildiyal M, Zamore PD (2009) Small silencing RNAs: an expanding universe. Nat Rev Genet 10:94–108
4. Klattenhoff C, Theurkauf W (2008) Biogenesis and germline functions of piRNAs. Development 135:3–9
5. Houwing S, Kamminga LM, Berezikov E, Cronembold D, Girard A, van den Elst H, Filippov DV, Blaser H, Raz E, Moens CB, Plasterk RH, Hannon GJ, Draper BW, Ketting RF (2007) A role for Piwi and piRNAs in germ cell maintenance and transposon silencing in Zebrafish. Cell 129:69–82

6. Vagin VV, Sigova A, Li C, Seitz H, Gvozdev V, Zamore PD (2006) A distinct small RNA pathway silences selfish genetic elements in the germline. Science 313:320–324
7. Brennecke J, Aravin AA, Stark A, Dus M, Kellis M, Sachidanandam R, Hannon GJ (2007) Discrete small RNA-generating loci as master regulators of transposon activity in *Drosophila*. Cell 128:1089–1103
8. Kawaoka S, Hayashi N, Suzuki Y, Abe H, Sugano S, Tomari Y, Shimada T, Katsuma S (2009) The *Bombyx* ovary-derived cell line endogenously expresses PIWI/PIWI-interacting RNA complexes. RNA 15:1258–1264
9. Kawaoka S, Izumi N, Katsuma S, Tomari Y (2011) 3′ End formation of PIWI-interacting RNAs in vitro. Mol Cell 43:1015–1022
10. Kawaoka S, Minami K, Katsuma S, Mita K, Shimada T (2008) Developmentally synchronized expression of two *Bombyx mori* Piwi subfamily genes, *SIWI and BmAGO3* in germline cells. Biochem Biophys Res Commun 367:755–760
11. Horwich MD, Li C, Matranga C, Vagin V, Farley G, Wang P, Zamore PD (2007) The *Drosophila* RNA methyltransferase, DmHen1, modifies germline piRNAs and single-stranded siRNAs in RISC. Curr Biol 17:1265–1272
12. Kirino Y, Mourelatos Z (2007) Mouse Piwi-interacting RNAs are 2′-*O*-methylated at their 3′ termini. Nat Struct Mol Biol 14:347–348
13. Ohara T, Sakaguchi Y, Suzuki T, Ueda H, Miyauchi K, Suzuki T (2007) The 3′ termini of mouse Piwi-interacting RNAs are 2′-*O*-methylated. Nat Struct Mol Biol 14:349–350
14. Saito K, Sakaguchi Y, Suzuki T, Suzuki T, Siomi H, Siomi MC (2007) Pimet, the *Drosophila* homolog of HEN1, mediates 2′-*O*-methylation of Piwi-interacting RNAs at their 3′ ends. Genes Dev 21:1603–1608
15. Kawaoka S, Mitsutake H, Kiuchi T, Kobayashi M, Yoshikawa M, Suzuki Y, Sugano S, Shimada T, Kobayashi J, Tomari Y, Katsuma S (2012) A role for transcription from a piRNA cluster in de novo piRNA production. RNA 18:265–273
16. Gunawardane LS, Saito K, Nishida KM, Miyoshi K, Kawamura Y, Nagami T, Siomi H, Siomi MC (2007) A slicer-mediated mechanism for repeat-associated siRNA 5′ end formation in *Drosophila*. Science 315: 1587–1590
17. Nishida KM, Saito K, Mori T, Kawamura Y, Nagami-Okada T, Inagaki S, Siomi H, Siomi MC (2007) Gene silencing mechanisms mediated by Aubergine piRNA complexes in *Drosophila* male gonad. RNA 13:1911–1922
18. Saito K, Nishida KM, Mori T, Kawamura Y, Miyoshi K, Nagami T, Siomi H, Siomi MC (2006) Specific association of Piwi with rasiRNAs derived from retrotransposon and heterochromatic regions in the *Drosophila* genome. Genes Dev 20:2214–2222
19. De Fazio S, Bartonicek N, Di Giacomo M, Abreu-Goodger C, Sankar A, Funaya C, Antony C, Moreira PN, Enright AJ, O'Carroll D (2011) The endonuclease activity of Mili fuels piRNA amplification that silences LINE1 elements. Nature 480:259–263
20. Reuter M, Berninger P, Chuma S, Shah H, Hosokawa M, Funaya C, Antony C, Sachidanandam R, Pillai RS (2011) Miwi catalysis is required for piRNA amplification-independent LINE1 transposon silencing. Nature 480:264–267

Chapter 5

A Framework for piRNA Cluster Manipulation

Ivan Olovnikov, Adrien Le Thomas, and Alexei A. Aravin

Abstract

Piwi proteins and their small-RNA partners, piwi-interacting (pi)RNA, form a natural mechanism that prevents the deleterious activity of transposable elements in the germ line of metazoan species. The piRNA pathway relies on extended noncoding genomic regions, dubbed piRNA clusters, to produce long precursor transcripts that are subsequently processed into mature piRNAs. The large size and repetitive nature of piRNA clusters provide significant challenges for their dissection using common genetic tools. Here we describe an effective approach for manipulation of piRNA clusters using a combination of BAC recombineering in *E. coli* and phiC31-mediated transgenesis in *Drosophila*. Although the described approach is instrumental for manipulating piRNA clusters, it can also be implemented for other problems in functional genomics.

Key words piRNA, piRNA cluster, BAC, Recombineering, phiC31

1 Introduction

The Piwi-interacting (pi)RNA is a distinct class of 23–31 nt RNA expressed predominantly in the germ line of Metazoa species. piRNAs guide their protein partners, Piwi proteins, to mRNA of active transposable elements. Upon recognition of target RNA, Piwi proteins use their endonucleolytic activity to destroy transposon transcripts and ensure genetic stability of the germ line [1–3].

In contrast to other classes of small RNA, such as microRNA and siRNA, the sequences of piRNA are extremely diverse and their processing does not require Dicer endonuclease [4]. piRNAs are processed from longer precursor RNA molecules; however, piRNA precursors do not have double-stranded RNA regions that are characteristic for miRNA and siRNA precursors. In *Drosophila*, the main source of piRNAs are discrete genomic loci called piRNA clusters that are usually devoid of protein-coding genes and are strongly enriched in transposon remnants [5]. Variations of piRNA pathway such as somatic piRNA expressed in follicular cells of *Drosophila melanogaster* ovary and pachytene piRNA expressed

Mikiko C. Siomi (ed.), *PIWI-Interacting RNAs: Methods and Protocols*, Methods in Molecular Biology, vol. 1093, DOI 10.1007/978-1-62703-694-8_5, © Springer Science+Business Media, LLC 2014

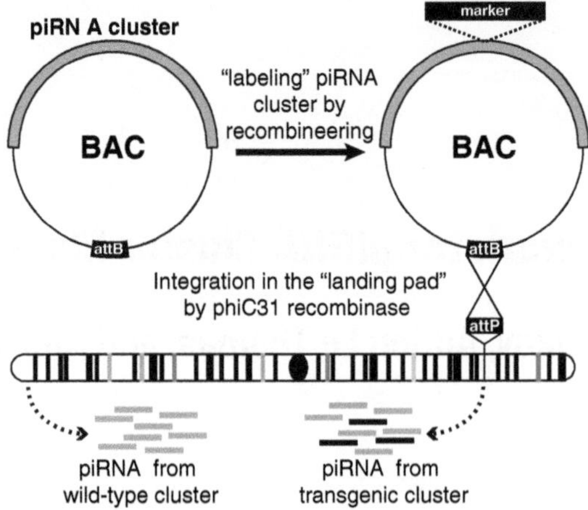

Fig. 1 Outline of the piRNA cluster modification strategy. A BAC clone containing full-length or fragment of piRNA cluster is tagged with a foreign sequence using bacterial recombineering. Modified BAC construct of large size can be inserted in a defined location within *Drosophila* genome by recombination between *attB* site in the BAC backbone and *attP* landing site in the genome. Recombination is carried out by phage phiC31 recombinase encoded in the genome of injection fly strain. If transgenic piRNA cluster is active in the ectopic position, artificial piRNA (*black*) are generated from the tag sequence and can be detected by piRNA profiling methods such as cloning and deep sequencing of piRNA libraries

during meiosis in mouse spermatocytes also rely on piRNA clusters as a major source of piRNAs [6–8]. piRNA clusters can span up to 200 kb and are believed to be transcribed as continuous units that generate long RNA precursors that are later processed into mature piRNA [5, 8].

In order to understand how transcription of piRNA cluster is regulated it is necessary to identify regulatory regions in piRNA clusters such as promoters and terminators. Bioinformatic studies failed to find primary or secondary sequence elements within piRNA precursors that would discriminate them from the rest of cellular transcripts and target for processing. Therefore, it is important to develop experimental approaches to dissect piRNA clusters in order to find critical elements that are necessary and sufficient for marking these regions as a source of piRNA. However, due to their big size and repetitive nature, manipulation of piRNA clusters in their native genome environment is challenging. Here we describe an approach to alter piRNA clusters cloned in bacterial artificial chromosomes (BAC) vectors and create transgenic flies that express modified piRNA clusters [9]. Because such transgenic piRNA cluster will be typically present on the background of native cluster in the fly genome, it is necessary to mark the transgenic cluster by insertion of an artificial tag sequence (Fig. 1). If the transgenic cluster is expressed and recognized for piRNA processing, the tag sequence is incorporated into the repertoire of cellular piRNAs that can be studied by cloning and deep sequencing of small RNA libraries. The number of piRNA from the tag serves as readout of the modified piRNA cluster activity.

The considerable size of piRNA clusters cloned in BAC prevents usage of traditional cloning techniques for their modification and P-element-mediated transformation for transgenesis. These problems can be addressed by a combination of two simple yet powerful techniques: BAC recombineering in bacteria and phiC31-recombinase-mediated transgenesis of *Drosophila* (Fig.1). The key resource for this approach is a P[acman] collection of BACs covering the *D. melanogaster* genome that can be accessed at http://www.pacmanfly.org [10]. P[acman] collection contains several clones corresponding to fragments or entire regions of several piRNA clusters. Importantly, the vector backbone used to create P[acman] clones allows efficient site-specific genomic integration of large BAC construct up to 130 kb using the bacteriophage phiC31 integrase [10, 11]. In principle, if the region of interest is not available in the P[acman] collection, other BACs can be modified by recombineering to make them amenable for phiC31-mediated transformation.

Recombineering in bacterial cells allows precise modification of virtually any sequence in the BAC, the only limitation being repeats [12–14]. The method is based on the homology-dependent recombination between BAC and linear DNA that is introduced into *E. coli* cells by electroporation. Recombination is assisted by the *Red* genes of bacteriophage λ encoded in a helper plasmid [12, 13, 15]. At minimum the linear DNA fragment should have two flanking homology arms of at least 50 bp that are used for recombination with BAC and antibiotic resistance gene that provides a selection marker to identify successful recombinants. It is possible to generate precise insertion, modification or deletion of a particular region in the BAC by changing regions used for homologous recombination and internal sequences in the linear DNA fragment.

In order to discriminate transgenic piRNA cluster from the native cluster of the same sequence we introduced an artificial sequence tag composed of a GFP sequence and a kanamycin resistance gene into the piRNA cluster cloned in the BAC [9]. The two components can be combined into one linear DNA using overlapping PCR. First, each fragment is amplified separately with one primer containing 50 nt homology arm and second containing ~20 nt that overlap with the second PCR product. Next, both PCR products are combined and amplified with two outer primers containing 50 nt homology arms that were used on the first step (Fig. 2). In one-step recombineering protocol, the antibiotic resistance gene that was used as a selection marker to identify successful recombinants remains in a modified BAC. Selection marker does not interfere with tagging of piRNA cluster, which can be performed with any sequence that is not present in *D. melanogaster* genome. Indeed, we found that artificial piRNAs are processed

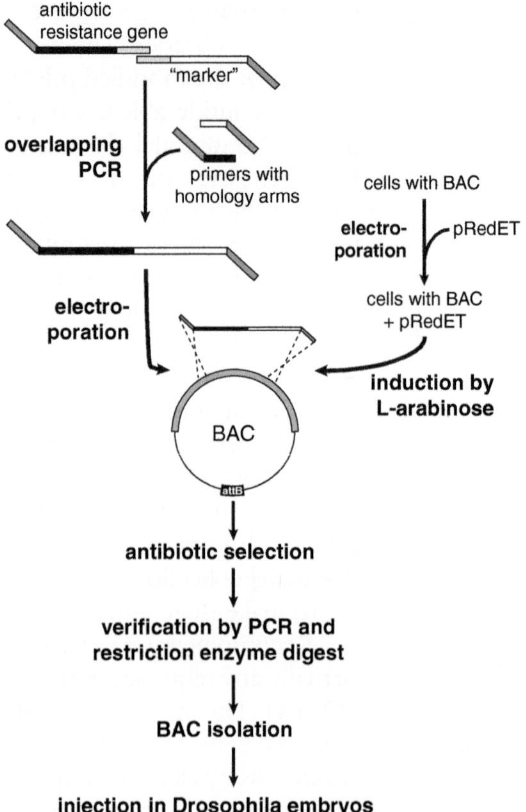

Fig. 2 The flow of BAC recombineering protocol. On the first step, a linear DNA fragment that contains antibiotic resistance gene and any additional sequence is generated by overlapping PCR. PCR product contains 50 bp homology arms on both ends, which correspond to sites of desired modification in the BAC. On the second step, cells carrying BAC of interest are transformed with pRedET plasmid, which encodes inducible genes necessary for efficient recombination between linear DNA and the BAC. Next, recombination is induced and cells are electroporated with a PCR product. Cells with integrated cassette are selected on appropriate antibiotic and integrity of BAC and correct insertion is verified by restriction digest and PCR

from both GFP and kanamycin genes in transgenic flies that carry insertion of modified piRNA cluster in the heterologous genomic locus [9]. This result indicates that transgenic clusters are active and artificial piRNAs of any sequence can be generated if the corresponding sequence is inserted in the cluster.

For further dissection of piRNA clusters it might be necessary to combine the tag sequence that generates artificial piRNAs with additional modifications of the cluster sequence. For example, the role of putative promoter in the piRNA cluster can be addressed by generating tagged cluster with deletion in the promoter region. If functional promoter is removed from the transgene, artificial piRNAs from the tag will not be produced in transgenic animals. For some experiments it might be desirable to generate "clean" modification of cluster sequence without insertion of selection marker. Such modification can be done using a technique that uses selection and counter-selection against inserted cassette in two sequential recombineering steps [16, 17].

After modification of piRNA cluster in the BAC clone by recombineering, modified BAC is introduced into fly genome by phiC31-mediated sequence-specific recombination between the *attB* site in the P[acman] backbone and the *attP* landing site in the fly genome. Fly strains with mapped landing sites integrated throughout the *D. melanoagster* genome were generated [11, 18–20]. Fixed position of a landing site provides a possibility to directly compare activity of different constructs in an identical genomic environment.

After transgenic animals are generated, expression of transgenic piRNA cluster is monitored by detection of artificial piRNAs generated from the tag sequence [9]. In principle it can be achieved by any method that allows sequence-specific detection of small RNAs such as Northern hybridization after separation of cellular RNA on 15 % denaturing PAAG gel. However, the cloning and sequencing of small RNA libraries either from total small RNA or RNA isolated from purified Piwi complexes provides by far the most deep and precise information about piRNA and should be used whenever possible [21]. Besides elimination of possible artifacts, profiling of piRNAs by deep sequencing provides the information about expression of transgenic piRNA cluster by comparing the amount of piRNAs derived from the tag to piRNAs generated from other clusters.

Overall, the combination of recombineering, phiC31-mediated transgenesis and analysis of cloned piRNA libraries provides excellent tool for dissection of functional elements in piRNA clusters. We expect that this approach will be instrumental in answering one of the most critical questions in the field: how piRNA precursors transcripts are discriminated from the rest of transcriptome and targeted for piRNA processing?

2 Materials

1. Template: plasmid DNA containing antibiotic resistance gene of choice (several different antibiotic resistance genes might be needed) and sequence that will be used as a tag (*lacZ* in the example described below).
2. Gel DNA Clean up kit (e.g., Zymoclean Gel DNA Recovery Kit, Zymo Research).
3. *E. coli* cells containing BAC of interest (backbone must contain phiC31 attB site; e.g., P[acman] collection BAC).

4. 20 ng/μl Plasmid pRedET (tet or amp) (Gene Bridges GmbH).
5. Sterile 10 % L-Arabinose in ddH_2O.
6. Electroporation device: Bio-Rad Gene Pulser II with Pulse Controller II module and compatible cuvettes (1 mm gap size; VWR).
7. PCR primers (*see* Subheadings 3 and 4).
8. BAC purification kit: Regular mini/midi-prep kit can be used for BACs up to 25 kb (e.g., Zyppy Plasmid Miniprep or Midiprep Kit, Zymo Research); BAC isolation kits should be used for large BACs (e.g., NucleoBond BAC 100, Macherey-Nagel).

3 Methods

3.1 Design the Strategy to Introduce Sequence Tag into piRNA Cluster

P[acman] collection is composed of two sets of BACs that differ in length: 21 and 83 kb on average. Several clones from these collections contain genomic regions corresponding to full-length and fragments of prominent piRNA clusters (*see* **Note 1**). The positions of P[acman] clones can be visualized in a genome browser available at http://bicoid.lbl.gov/cgi-bin/gbrowse/getbac/.

To tag a piRNA cluster for generation of artificial piRNAs, a 100 bp non-repetitive stretch with 40–60 % GC ratio needs to be chosen within a BAC of interest. As generation of piRNA is not uniform along the cluster, it is advisable to select a location with high density of piRNA. Selection of suitable BAC and site for insertion is an important step and should be done carefully because of the repetitive nature of piRNA clusters in *Drosophila*. Some BACs are unstable upon induction of recombination because of repeats.

In BAC recombineering a linear DNA fragment usually generated by PCR can be used for deletion, insertion or mutagenesis of a particular region in the BAC. Independently of the type of modification, the PCR product must contain antibiotic resistance gene to allow selection of clones after recombination. In addition, a PCR product must contain 50 bp homology arms on each end that are identical to the flanks of the target sequence in the BAC. For deletion, antibiotic resistance gene is amplified with primers containing 50 nt homology arms flanking the region to be deleted. If seamless deletion without leaving resistance gene is required, refer to the optional section in the end of this protocol.

Insertion of extra sequence in the BAC containing piRNA cluster sequence can be made in the same manner as deletion: the antibiotic resistance gene is inserted using homology recombination and its sequence will serve as a marker from which unique piRNA are produced (Fig. 1). An additional sequence can be added to the insertion cassette, which potentially might allow functional readout of artificial piRNA generation by measuring silencing of cognate reporter gene located elsewhere in the genome.

3.2 Preparation of PCR Product for Homologous Recombination

Here we describe BAC recombineering on an example of insertion kanR-lacZ cassette into P[acman] BAC containing piRNA cluster (clone CH322-175K12 corresponding to chromosome X positions 21390119–21411295). The insertion is made at the site that corresponds to genomic position chrX: 21397502 (*see* **Note 2**).

To generate the linear DNA fragment that contains both kanR and lacZ genes, two PCR products need to be joined by overlap PCR. First, each fragment is amplified individually. PCR products must contain at least 20 bp of identical sequence on the end that will be joined, so that each PCR product will serve as a primer in overlap PCR. On the second step, two PCR products are mixed in equimolar ratio and amplified with flanking primers, which contain 50 nt homology arms (Fig. 2).

1. PCR amplify kanR gene using primers 1-fwd and 1-rev (almost any kanR-containing plasmid can be used, e.g., pDONR201, Invitrogen) and lacZ gene using primers 2-fwd and 2-rev (on any vector carrying lacZ gene, e.g., pCaSpeR-AUG-bGal). Homology arms that will be later used for integration into BAC are underlined. Note that primer 2-fwd contains additional 20 nucleotides that are complementary to primer 1-rev and will be used on **step 2** for overlapping PCR (marked by bold). Purify PCR products by excision from agarose gel and column purification (*see* **Notes 3–5**).

   Primer 1-fwd

   <u>TCCAAAGACTAAAAGGATCAATTGACAATATAATGGAGATCTATGCTTGG</u>GCAGCTCTGGCCCGTGTCTC

   Primer 1-rev

   GTCTGACGCTCAGTGGAACG

   Primer 2-fwd

   CGTTCCACTGAGCGTCAGACTTTGCCTGGTTTCCGGCACCA

   Primer 2-rev

   <u>ATCTTACTTTGACCACTGTTAATCCATGCTAAGGTAATCGTGGAATCACT</u>AAGGGCTGGTCTTCATCCAC

2. Perform overlapping PCR with two purified PCR products obtained on **step 1**. Add primers 1-fwd and 2-rev and equimolar amounts of each PCR product to reaction: 10 ng of PCR 1 (~1,100 bp) and 13 ng of PCR 2 (~1,400 bp). Purify PCR product by excision from agarose gel and column purification. Elute in small volume of water and adjust concentration of DNA to ~100 ng/μl with water (*see* **Note 3**).

3.3 Homologous Recombination of PCR Product and BAC

Day 1: Preparation of cells culture with BAC

1. Start overnight culture (1–5 ml of LB medium) of cells containing BAC of interest. In case of P[acman] BACs add 12.5 μg/ml chloramphenicol for selection.

Day 2: Transformation with pRedET plasmid

1. Prepare ice-cold sterile deionized water and electroporation cuvettes. Precool centrifuge to 4 °C.
2. Inoculate 30 μl of fresh overnight culture into 1.4 ml of LB medium supplemented with 12.5 μg/ml of chloramphenicol and incubate at 37 °C for 2–3 h with vigorous shaking. **Steps 3** and **4** should be performed in cold room (*see* **Note 6**).
3. Centrifuge cells in the refrigerated (4 °C) centrifuge for 30 s at 11,000 × *g*. Discard supernatant and resuspend cells in 1 ml of ice-cold sterile water. Mix by gentle pipetting.
4. Repeat washing **step 3**. Remove supernatant and resuspend cells in 30 μl of ice-cold deionized water. Add 1 μl of pRedET plasmid (20 ng/μl) and transfer cells into cold electroporation cuvette (*see* **Note 7**).
5. Electroporate cells at 1.35 kV, 10 μF, 600 Ω. Pulse should be in a range of 5–6 ms.
6. Immediately resuspend cells in 1 ml of LB medium without antibiotics and transfer in a 1.7-ml eppendorf tube. Shake the tube at 30 °C for 70 min (*see* **Note 7**).
7. Centrifuge cells at 11,000 × *g* for 30 s, remove supernatant, and resuspend the pellet in 100 μl of LB medium. Plate cells on agar plate containing 12.5 μg/ml chloramphenicol and 3 μg/ml tetracycline (for selection of pRedET (tet) plasmid; if other vector is used, second antibiotic should be changed accordingly). Incubate the plates at 30 °C overnight.

Day 3: Preparation of cell culture with BAC and pRedET plasmid

1. Inoculate several colonies (up to 5) of cells containing BAC and pRedET plasmid in a single 5 ml volume of LB medium supplemented with 12.5 μg/ml chloramphenicol and 3 μg/ml tetracycline. Incubate overnight at 30 °C.

Day 4: Insertion of PCR product into BAC

1. Prepare ice-cold sterile deionized water and electroporation cuvettes. Precool centrifuge to 4 °C.
2. Inoculate 30 μl of fresh overnight culture into two tubes with 1.4 ml of LB medium with chloramphenicol and tetracycline and incubate at 30 °C for 2 h with vigorous shaking.
3. Add 50 μl of 10 % L-Arabinose to one of the tubes to induce expression of genes necessary for mediate recombination; second tube without L-Arabinose serves as negative control. Incubate both cultures for 1 h at 37 °C (*see* **Notes 7** and **8**).
4. Proceed to electroporation as described in **steps 3–7** of Day 2. On **step 4** use by 2 μl of purified PCR product (~100 ng/μl) for electroporation of both control and experiment cells.

On **step 6** incubate cells at 37 °C instead of 30 °C (*see* **Note 7**). On **step 7** plate cells on LB plates with chloramphenicol and kanamycin (or other antibiotic dependent on resistance gene present in the insertion cassette) but without tetracycline. Incubate plates at 37 °C. A significant difference in colony number should be observed between induced and control plates.

Days 5–7: Verification of recombineering products

1. Inoculate several colonies with recombinant BAC in 5–20 ml volume of LB medium supplemented with chloramphenicol and kanamycin. Grow overnight at 37 °C. In case of P[acman] BACs, copy number can be increased by adding L-arabinose to the concentration of 0.1 %.
2. Save glycerol stocks of each clone (mix 300 μl of 50 % sterile glycerol and 700 μl of overnight culture; store at −80 °C) and prepare DNA using regular plasmid DNA isolation kit (enough for verification purposes).
3. Verify clones by restriction digest and PCR. For restriction digest perform several independent reactions with different enzymes for each BAC clone. Choose enzymes that cut both within and outside of insertion. Always set up control digests of the unmodified original BAC. Typically, more than a dozen of digest products are resolved on a long 0.7–1 % agarose gel. If large fragments are examined, run 0.7 % agarose gel in a cold room at low voltage. For PCR verification choose primer pairs that overlap left and right junctions of inserted sequence. Make sure that primers lie beyond the 50 bp homology regions. Include unmodified BAC as a control (*see* **Note 9**).
4. Choose correct clone and use glycerol stock to inoculate large culture for preparative purpose. Prepare BAC DNA using either regular midiprep kit (for BACs up to 25 kb) or BAC isolation kit (for larger BACs). Elute only with double deionized sterile water. Keep concentration in a range of 0.5–1 μg/ml.
5. Submit DNA for embryo injection. If injection is performed by commercial company, refer to its requirements regarding amount and concentration of DNA. Select the attP landing site for integration. Typically, the landing site that gives most transformants is preferred (*see* **Note 10**).
6. After fly stock carrying transgenic piRNA cluster is established proceed with preparation of small RNA libraries for deep sequencing [21]. Small RNA can be cloned from total ovarian RNA or from RNA isolated from distinct immunopurified Piwi complexes. Sequencing of small RNA from original stock without BAC transgene is recommended as a control.

3.4 Optional: Seamless BAC Modification by Two-Step Recombineering

Two-step recombineering can be used if the presence of the antibiotic resistance gene in the modified BAC interferes with the experimental design, e.g., when a precise deletion or point mutagenesis is required. Briefly, the first step of recombineering is performed as described and clones with correct insertions are selected. Next, the second recombineering step is used to remove the cassette inserted on the first step and replace it with desired fragment that can be of any sequence. For two-step recombineering, in addition to the standard antibiotic resistance gene, the insertion cassette should contain a counter-selection gene. This gene is usually either *sacB*, which confers sensitivity to sucrose, or *rpsL*, which confers sensitivity to streptomycin. For example, to delete the entire cassette without introduction any sequences, cells should be electroporated with a 100 nt long oligonucleotide which consists of two homology arms used on the first step. If the *rpsL* gene was present in the cassette, plating on streptomycin would select for cells in which cassette excision has occurred. Commercially available Counter-Selection BAC Modification Kit (#K002, Gene Bridges GmbH) contains reagents and detailed instructions for the two-step recombineering procedure.

4 Notes

1. Although BAC ends of the P[acman] collection were mapped to genome we prefer to confirm them by sequencing with SP6 and T7 primers.
2. UCSC Genome Browser (http://genome.ucsc.edu/cgi-bin/hgGateway) is an indispensable tool for visualization and sequence retrieval of genomic data. Built-in BLAT tool allows simultaneous search of multiple (up to 25) sequences, therefore features of interest can be highlighted. This, combined with the possibility to add custom tracks, such as piRNA density, allows selection of the region for BAC modification in a convenient way. piRNA analysis and generation of tracks for UCSC genome browser can be done using Galaxy tools (https://main.g2.bx.psu.edu).
3. Any thermostable polymerase with the corresponding buffer can be used for amplification. Refer to manufacturer instructions for PCR conditions. We routinely use KOD Hot Start DNA Polymerase from Novagen. Supplementing reaction with 1 M betaine and decreasing annealing temperature to 5 °C below the calculated Tm often helps to increase yields and specificity, particularly during overlapping PCR. For gel purification we use Zymoclean Gel DNA Recovery Kit (Zymo Research) because it allows elution in small volume

(6 μl) and therefore provides high concentration of DNA for electroporation.

4. We routinely use primers with homology arms for first PCRs and the same primers later for overlapping PCR. Alternatively, short primers without homology arms can be used on the first PCR step and longer primers with homology arms for overlapping PCR.
5. It is important to verify that PCR products do not share any homology with the BAC insert or vector backbone except for the region of desired recombination. Any sequences of plasmid or transposon origin present in the PCR product should be examined carefully (remember that recombination can also occur between BAC backbone and the PCR product if they share homology).
6. We noticed that work in cold room at these steps with all plastic and solutions cooled to 4 °C and keeping cells on ice at all times is critical for successful transformation.
7. For recombineering we use pRedET (tet) plasmid (Gene Bridges GmbH), which encodes L-arabinose inducible phage-derived genes needed for homology-dependent recombination. Propagation of pRedET plasmid in *E. coli* cells is temperature-sensitive: the cells can be cured from the plasmid by transfer from 30 °C (permissive temperature) to 37 °C. Accordingly, until recombination is induced by addition of L-arabinose, the cells should be maintained at 30 °C or below. Simultaneous curing of plasmid and induction of genes encoded on it by shifting to 37 °C prevents overexpression of recombination genes that can be toxic for cells and cause unwanted BAC rearrangements.
8. Copy number of P[acman] BACs in *E. coli* cells is induced by L-arabinose [11]. At the same time L-arabinose induces expression of recombination genes from pRedET plasmid. Although copy number of BAC should be minimized during recombination step we did not observe any interference of these two seemingly antagonistic processes.
9. We recommend creating in silico maps of entire constructs, including vector backbone. This greatly simplifies design of restriction digests and PCR tests and is helpful during analysis of small RNA data from transgenic animals.
10. We use services of BestGene, Inc. (http://www.thebestgene.com) for *Drosophila* transformation. BestGene has a vast selection of fly stocks with different attP landing sites. For BAC transgenesis we prefer stock number 9750 (landing site is located in chromosome position 65B2), which consistently yields sufficient number of transformants. For large BAC it is recommended to inject at least 400 embryos.

Acknowledgments

We thank members of the Aravin lab for helpful discussion and comments on the manuscript. We are particularly thankful to Alexandre Webster for comments and editing. I.O. is a CEMI (Center for Environmental Microbiology Interactions) fellow at Caltech. This work was supported by grants from the National Institutes of Health (R01 GM097363, R00 HD057233, and DP2 OD007371A) and the Searle Scholar Award to A.A.A.

References

1. Khurana JS, Theurkauf W (2010) piRNAs, transposon silencing, and *Drosophila* germline development. J Cell Biol 191:905
2. Senti KA, Brennecke J (2010) The piRNA pathway: a fly's perspective on the guardian of the genome. Trends Genet 26:499
3. Siomi MC, Miyoshi T, Siomi H (2010) piRNA-mediated silencing in *Drosophila* germlines. Semin Cell Dev Biol 21:754
4. Vagin VV, Sigova A, Li C, Seitz H, Gvozdev V, Zamore PD (2006) A distinct small RNA pathway silences selfish genetic elements in the germline. Science 313:320
5. Brennecke J, Aravin AA, Stark A, Dus M, Kellis M, Sachidanandam R, Hannon GJ (2007) Discrete small RNA-generating loci as master regulators of transposon activity in *Drosophila*. Cell 128:1089
6. Malone CD, Brennecke J, Dus M, Stark A, McCombie WR, Sachidanandam R, Hannon GJ (2009) Specialized piRNA pathways act in germline and somatic tissues of the *Drosophila* ovary. Cell 137:522
7. Li C, Vagin VV, Lee S, Xu J, Ma S, Xi H, Seitz H, Horwich MD, Syrzycka M, Honda BM, Kittler EL, Zapp ML, Klattenhoff C, Schulz N, Theurkauf WE, Weng Z, Zamore PD (2009) Collapse of germline piRNAs in the absence of Argonaute3 reveals somatic piRNAs in flies. Cell 137:509
8. Aravin A, Gaidatzis D, Pfeffer S, Lagos-Quintana M, Landgraf P, Iovino N, Morris P, Brownstein MJ, Kuramochi-Miyagawa S, Nakano T, Chien M, Russo JJ, Ju J, Sheridan R, Sander C, Zavolan M, Tuschl T (2006) A novel class of small RNAs bind to MILI protein in mouse testes. Nature 442:203
9. Muerdter F, Olovnikov I, Molaro A, Rozhkov NV, Czech B, Gordon A, Hannon GJ, Aravin AA (2012) Production of artificial piRNAs in flies and mice. RNA 18:42
10. Venken KJ, Carlson JW, Schulze KL, Pan H, He Y, Spokony R, Wan KH, Koriabine M, de Jong PJ, White KP, Bellen HJ, Hoskins RA (2009) Versatile P[acman] BAC libraries for transgenesis studies in *Drosophila* melanogaster. Nat Methods 6:431
11. Venken KJ, He YR, Hoskins A, Bellen HJ (2006) P[acman]: a BAC transgenic platform for targeted insertion of large DNA fragments in D. melanogaster. Science 314:1747
12. Muyrers JP, Zhang Y, Stewart AF (2001) Techniques: recombinogenic engineering—new options for cloning and manipulating DNA. Trends Biochem Sci 26:325
13. Zhang Y, Buchholz F, Muyrers JP, Stewart AF (1998) A new logic for DNA engineering using recombination in Escherichia coli. Nat Genet 20:123
14. Datsenko KA, Wanner BL (2000) One-step inactivation of chromosomal genes in Escherichia coli K-12 using PCR products. Proc Natl Acad Sci U S A 97:6640
15. http://www.genebridges.com/gb/red_et_principles.php
16. Muyrers JP, Zhang Y, Benes V, Testa G, Ansorge W, Stewart AF (2000) Point mutation of bacterial artificial chromosomes by ET recombination. EMBO Rep 1:239
17. http://www.genebridges.com/gb/details.php?prod_id=K002&main_group=red
18. Bischof J, Maeda RK, Hediger M, Karch F, Basler K (2007) An optimized transgenesis system for *Drosophila* using germ-line-specific phiC31 integrases. Proc Natl Acad Sci U S A 104:3312
19. Markstein M, Pitsouli C, Villalta C, Celniker SE, Perrimon N (2008) Exploiting position effects and the gypsy retrovirus insulator to engineer precisely expressed transgenes. Nat Genet 40:476
20. http://www.thebestgene.com/PhiC31InfoPage.do
21. Malone C, Brennecke J, Czech B, Aravin A, Hannon GJ (2012) Preparation of small RNA libraries for high-throughput sequencing. Cold Spring Harb Protoc 2012:1067

Chapter 6

Biochemical and Mass Spectrometric Analysis of 3'-End Methylation of piRNAs

Takeo Suzuki, Kenjyo Miyauchi, Yuriko Sakaguchi, and Tsutomu Suzuki

Abstract

Piwi-interacting RNAs (piRNA) are fully modified by 2′-*O*-methylation at their 3′-termini. This terminal methylation is required to prevent 3′-nucleotide addition, which serves as a tag for destabilization. In this chapter, we describe biochemical and mass spectrometric analyses of 2′-*O*-methylation at 3′-termini of piRNAs.

Key words Posttranscriptional modification, 2′-*O*-methylation, HEN1, RNA preparation, AGPC method, Periodate oxidation, Northern blotting, Mass spectrometry, LC/MS

1 Introduction

Piwi-interacting RNAs (piRNA) are small noncoding RNAs with lengths of 26–31 nucleotides, and they are abundantly expressed in the germline in animals. Biogenesis and function of piRNAs is associated with normal germ cell development through silencing retrotransposons and repetitive elements [1–3]. One of the characteristic structural features of piRNAs is the presence of posttranscriptional modification. We and the Mourelatos laboratory reported that mouse piRNAs have 2′-*O*-methylation at their 3′-termini [4, 5]. The same modification was found in zebrafish and *Drosophila* piRNAs [6–9]. The 3′-end 2′-*O*-methyl modification was previously found in *Arabidopsis* miRNAs and siRNAs [10, 11], whereas 3′-ends of animal miRNAs remain unmodified. Hua enhancer 1 (HEN1) was identified as an RNA methyltransferase responsible for the 3′-end 2′-*O*-methylation of plant miRNAs [10]. Plant HEN1 has a double-stranded RNA binding domain (dsRBD) and employs an miRNA/miRNA* duplex as a substrate for 2′-*O*-methylation [12]. In plants lacking HEN1, miRNAs undergo 3′-uridylation mediated by HESO1 and are destabilized, resulting in pleiotropic phenotypes [13]. Hence, the 3′-end

Mikiko C. Siomi (ed.), *PIWI-Interacting RNAs: Methods and Protocols*, Methods in Molecular Biology, vol. 1093,
DOI 10.1007/978-1-62703-694-8_6, © Springer Science+Business Media, LLC 2014

2′-*O*-methyl group is required to maintain normal maturation and stability of plant miRNAs. A bioinformatics survey suggested that HEN1 family genes are widely distributed in eukaryotes and bacteria [14]. Animal HEN1 is strongly expressed in gonads, and it colocalized with Piwi protein in nuage [15–17]. Biochemical analyses revealed that animal HEN1 can catalyze 2′-*O*-methylation at 3′-ends of piRNAs and their counterparts [6, 9, 15–20]. Unlike plant HEN1, animal HEN1 lacks dsRBD and acts on single-stranded RNA. In an in vitro system for loading piRNA precursor into Piwi protein, the 2′-*O*-methylation tightly couples with 3′-to-5′ exonucleolytic trimming of piRNA precursor on Piwi protein [20]. In zebrafish, HEN1 is essential for maintaining a female germ line, whereas it is dispensable in the testis [16]. However, testis piRNAs become 3′-uridylated and adenylated in the *hen1* mutant [16]. In *Drosophila*, in the absence of the 3′-end methylation, the length and abundance of piRNAs are decreased, and piRNA function is perturbed [6]. In addition, 2′-*O*-methylation by DmHen1 prevents addition and trimming of 3′-nucleotides in Argonaute2-associated siRNAs [21]. Taken together, the 3′-end 2′-*O*-methylation regulates the physiological role of piRNA via protecting the 3′-ends from terminal nucleotide addition and/or exonucleolytic degradation.

In this chapter, we describe RNA preparation from mouse testis, biochemical detection of 3′-end 2′-*O*-methylation of piRNAs by periodate oxidation and Northern blotting, and direct observation of 3′-terminal 2′-*O*-methyl nucleosides of piRNAs by liquid chromatography-mass spectrometric (LC/MS) analysis.

2 Materials

2.1 RNA Preparation and Gel Purification of the piRNA Fraction

1. Mouse testis (frozen and stored at −80 °C, 0.1 g/sample).
2. RNase-free water: DEPC-treated water or milli-Q water (we use milli-Q water directly for RNA work).
3. Denaturing solution (solution D): 4 M guanidinium thiocyanate, 25 mM sodium citrate (pH 7.0), 0.5 % (w/v) *N*-lauroylsarcosine (Sarkosyl), and 0.1 M 2-mercaptoethanol; 2-mercaptoethanol should be added just before use.
4. 2 M NaOAc (pH 4.0).
5. Water-saturated phenol.
6. Chloroform:isoamyl alcohol (49:1 v/v; CIA).
7. 3 M NaOAc (pH 5.2).
8. Ethanol.
9. TRI Reagent (Sigma) or other acid-guanidinium-phenol based reagents, such as TRIzol (Life Technologies), TriPure (Roche), ISOGEN (Nippon Gene), or QIAzol (Qiagen).

10. 80 % ethanol.
11. Amicon Ultra-15 centrifugal filter unit with Ultracel-100 membrane (100 kDa NMWL, Millipore).
12. 1×TBE (10.8 g/L Tris, 5.5 g/L boric acid, 4 mM EDTA).
13. Denaturing acrylamide gel solution: 15 % acrylamide/*N*,*N*′-methylenebisacrylamide (19:1), 1×TBE, 7 M urea.
14. *N*,*N*,*N*′,*N*′-tetramethylethylenediamine (TEMED).
15. 40 % ammonium persulfate (APS).
16. 2×Loading dye: 8 M urea, 1×TBE, 30 % glycerol, 0.025 % bromophenol blue, and 0.05 % xylene cyanol.
17. Small RNA marker: DynaMarker small RNA II (BioDynamics Laboratory).
18. SYBR Gold (Life Technologies).
19. Liquid nitrogen.
20. POLYTRON homogenizer (Kinematica AG).
21. Wooden mallet.

2.2 Periodate Oxidation and Northern Hybridization

1. Sodium periodate.
2. Lysine monohydrochloride.
3. Hybond-N$^+$ (GE Healthcare).
4. 3 MM CHR filter paper (Whatman).
5. Trans-blot SD semi-dry transfer cell (Bio-Rad).
6. Optikinase (Affymetrix).
7. [γ-^{32}P] ATP (MP Biomedicals).
8. UV crosslinker (UVP, CL-1000).
9. ULTRAhyb-Oligo (Ambion).
10. 2×SSC (0.3 M NaCl, 30 mM sodium citrate).

2.3 Mass Spectrometric Analysis

1. RNase T_2 (Pharmacia).
2. RNase T_1 (Epicentre).
3. RNase A (Ambion).

   For the following reagents (4–9), LC/MS grade (at least analytical grade) is recommended for LC/MS analysis:
4. Ammonium acetate.
5. Acetonitrile.
6. 1,1,1,3,3,3-Hexafluoro-2-propanol (HFIP).
7. Methanol.
8. Triethylamine.
9. Acetic acid

The LC/MS systems that we used for piRNA and general RNA MS analyses are described in detail in published literature [5, 22–24]. Any similar instruments can be adapted for LC/MS systems.

10. A micro-flow LC/MS system composed of an HP1100 liquid chromatography system (Agilent) and LCQ-Advantage (Thermo Fisher Scientific) with an electrospray ionization source.
11. An ODS reversed-phase column (Inertsil ODS-3, ID 2.1 × 250 mm, GL Sciences) with a pre-column cartridge (Inertsil ODS-3, ID 3 × 10 mm, GL Sciences) for the micro-flow separation.
12. A nano-flow LC/MS system, composed of a splitless nano-HPLC system DiNa (KYA Technologies) with a Nanovolume Valve (Valco Instruments) and a QSTAR XL (Applied Biosystems), or LTQ Orbitrap XL (Thermo Fisher Scientific), equipped with a nano-electrospray interface.
13. An ODS capillary column (HiQ sil, ID 0.15 × 50 mm, KYA Technologies) and an ODS trapping column (HiQ sil, ID 0.5 × 0.1 mm, KYA Technologies) for the nano-flow separation.

3 Methods

3.1 RNA Extraction from Testes by AGPC Method

1. Prepare and chill the POLYTRON homogenizer and reagents in the cold room.
2. Quickly weigh frozen testes. Do not let the tissues thaw.
3. Shape a freezer bag (or hybridization bag, OHP film for handwriting) to form an appropriate size of bag with a bag sealer.
4. Pour some liquid N_2 into a styrofoam box.
5. Transfer the testes into the freezer bag, and immerse it in liquid N_2 (*see* **Note 1**).
6. Quickly crush the frozen testes in the bag with a wooden mallet to make it into small pieces. Do not let the tissues thaw. If the tissue begins to melt, immerse it again in liquid N_2.
7. Cut off a part of the bag. Transfer the frozen powder of testis into an appropriate size polypropylene tube (50 mL centrifuge tube or 0.5–1 L centrifuge bottle).
8. Add 1–2 mL solution D per 0.1 g testis to the tube.
9. Disrupt the tissue with the POLYTRON homogenizer. Break down the lumps of the tissue by handling of the homogenizer. After all lumps are broken, continue the homogenization until the homogenate becomes creamy.

10. Add 0.1 volume (per volume of solution D) of 2 M NaOAc (pH 4.0) while washing the generator shaft with this buffer.
11. Add an equal volume (per volume of solution D) of water-saturated phenol, and shake the tube vigorously.
12. Leave the tube for approximately 5 min at 4 °C or room temperature with occasional shaking.
13. Centrifuge at 4,500–9,000 × *g* (maximum speed allowed for the centrifuge) for 30 min at 4 °C to remove sticky cell debris (*see* **Note 2**).
14. Transfer the entire supernatant to a clean tube. Do not take the sticky pellet of debris.
15. Add 0.2 volume of CIA (*see* **Note 3**) and vortex vigorously.
16. Let the tubes stand for 15 min at 4 °C or room temperature.
17. Centrifuge at 4,500–9,000 × *g* for 30 min at 4 °C.
18. Carefully transfer the upper aqueous phase, which contains mostly RNA, to a clean tube using a pipette (*see* **Note 4**).
19. Add 0.2 volume of chloroform to the aqueous phase, and shake vigorously (*see* **Note 5**).
20. Transfer the upper aqueous phase to a clean tube.
21. Add 0.1 volume of 3 M NaOAc (pH 5.2) and 2.5 volumes of ethanol (*see* **Note 6**).
22. Incubate the samples for at least 1–2 h at −80 °C.
23. Centrifuge at 4,500–9,000 × *g* for 30 min at 4 °C and discard the supernatant. The pellet of RNA, which usually contains other contaminants, should be formed.
24. Dissolve the RNA pellet in RNase-free water (*see* **Note 7**).
25. Add 0.2 volume of chloroform to the aqueous solution, and shake vigorously (*see* **Note 8**).
26. Centrifuge at 4,500–9,000 × *g* for 30 min at 4 °C.
27. Transfer the upper aqueous phase to a clean tube.
28. Add an equal volume of TRI reagent, and vortex vigorously (*see* **Note 9**).
29. Add 0.2 volume of chloroform to the aqueous phase, and shake vigorously.
30. Centrifuge at 4,500–9,000 × *g* for 30 min at 4 °C.
31. Transfer the upper aqueous phase to a clean tube.
32. Add 0.1 volume of 3 M NaOAc (pH 5.2) and 2.5 volumes of ethanol.
33. Incubate the samples for at least 1–2 h at −80 °C.
34. Centrifuge at 4,500–9,000 × *g* for 30 min at 4 °C and discard the supernatant.

35. Briefly centrifuge, and remove the residual supernatant completely.
36. Dissolve the RNA pellet in RNase-free water (*see* **Note 10**).
37. Add 0.1 volume of 3 M NaOAc (pH 5.2) and 2.5 volumes of ethanol.
38. Incubate the samples for at least 1–2 h at −80 °C.
39. Centrifuge at 4,500–9,000 × *g* for 30 min at 4 °C and discard the supernatant.
40. Briefly centrifuge and remove the residual supernatant completely.
41. Rinse the pellet with a sufficient volume of 80 % ethanol (*see* **Note 11**), centrifuge at maximum speed for 20–30 min, and discard the supernatant.
42. Briefly centrifuge and remove the residual supernatant completely.
43. Air-dry or vacuum-dry the RNA pellet briefly to remove residual ethanol (*see* **Note 12**).
44. Dissolve the RNA pellet in RNase-free water.
45. Evaluate the quantity and purity of the extracted RNA by measurement of ultraviolet absorption.

3.2 Fractionation of Small RNA by Ultrafiltration

1. Transfer the total RNA solution (up to 12 mL) into the Amicon Ultra-15 centrifugal filter units (100 kDa cut off).
2. Centrifuge at 5,000 × *g* for more than 45 min at 8 °C (*see* **Note 13**).
3. Transfer the filtrate to a clean tube.
4. Add RNase-free water (an equal volume of the sample) to the filter units and dissolve the residue by pipetting.
5. Centrifuge again at 5,000 × *g* for more than 45 min at 8 °C.
6. Transfer the filtrate to the tube.
7. Add 0.1 volume of 3 M NaOAc (pH 5.2) and three volumes of ethanol (*see* **Note 14**) and mix well.
8. Incubate the samples overnight at −80 °C.
9. Centrifuge at maximum speed for 30 min at 4 °C and discard the supernatant.
10. If the solution contains large amounts of low molecular weight compounds, the pellet should be dissolved with RNase-free water again and repeat the step of ethanol precipitation.
11. Rinse the pellet with a sufficient volume of 80 % ethanol, centrifuge at maximum speed for 20–30 min, and discard the supernatant.
12. Briefly centrifuge and remove the residual supernatant completely.

13. Vacuum-dry the RNA pellet.
14. Dissolve the RNA pellet in RNase-free water.

3.3 Preparation of piRNA Fraction

1. Prepare a 15 % acrylamide gel (19:1) containing 7 M urea with a comb with a broad well.
2. Mix the small RNA fraction (up to 400 μg per gel) with 2×loading dye.
3. Heat samples at 70–80 °C for 2 min.
4. Wash out the urea in the well, and load the samples and markers.
5. Run the gel at 250–300 V until the xylene cyanol dye migrates about 5/8 down the gel.
6. Cut the gel in half to separate the portion that contains the very high concentration of tRNA.
7. Stain the lower portion of the gel with SYBR Gold.
8. Cut out the gel corresponding to the piRNA band (approximately 30 nucleotides) with a clean razor blade, and put it into the tube.
9. Elute the piRNA fraction from the gel by the crush and soak method.
10. Precipitate the RNA by ethanol precipitation. Use 3–4 volumes of ethanol and incubate overnight at −80 °C.

3.4 Periodate Oxidation

1. The small RNA fraction (approximately 20 μg, *see* Subheading 3.2) in 100 μL solution of 10 mM sodium periodate is incubated at 0 °C for 40 min in dark.
2. Add 0.1 volume of 3 M NaOAc (pH 5.2) and 2.5 volumes of ethanol.
3. Incubate the samples for at least 1–2 h at −80 °C.
4. Centrifuge at 20,000×*g* for 15 min at 4 °C and discard the supernatant.
5. Briefly centrifuge and remove the residual supernatant completely.
6. Dissolve the RNA pellet in 100 μL of RNase-free water and repeat **steps 2–5**, Subheading 3.4.
7. Rinse the pellet with a sufficient volume of 80 % ethanol, centrifuge at maximum speed for 15 min, and discard the supernatant.
8. Briefly centrifuge and remove the residual supernatant completely.
9. Briefly air-dry or vacuum-dry the RNA pellet to remove residual ethanol.

10. Dissolve the RNA pellet in 100 μL of RNase-free water.
11. Add an equal volume of 2 M lysine-HCl (prepare immediately prior to use) and incubate at 45 °C for 90 min (β-elimination).
12. Repeat **steps 2–9** in Subheading 3.4 (two times ethanol precipitation and 80 % ethanol rinse).
13. Dissolve the RNA pellet in 50 μL of RNase-free water.

3.5 Northern Hybridization of the piRNA Fraction

1. Prepare a 15 % denaturing acrylamide gel with a 10- to 15-well gel comb. The size of the gel plates is 20 cm × 20 cm × 1 mm.
2. Mix 2 μg of oxidized (5 μL from Subheading 3.4) and non-treated small RNA fraction with an equal volume of 2× loading dye.
3. Wash out the urea in the wells of the gel (in the electrophoresis assembly).
4. Load the samples and RNA markers into each well, and run the gel at 800–1,000 V until the xylene cyanol dye migrates 12–15 cm from the well.
5. Immediately after electrophoresis, detach the gel from the assembly and remove the gel plates.
6. Stain the lower portion of the gel with SYBR Gold and check the position of the marker.
7. Cut the gel in such a way as to contain the range of 15–40 nucleotide length (10 × 10 to 15 × 15 cm^2; smaller is preferable for handling). Cut a sheet of Hybond-N^+ nylon membrane and 10 pieces of 3 MM CHR filter paper to the size of the excised gel.
8. Soak the gel, membrane, and filter paper with 1× TBE and avoid drying them.
9. Gently place the gel, membrane, and filter paper in a semi-dry electroblot apparatus (Trans-Blot SD Semi-Dry Transfer Cell). The order of placement is anode, 5 pieces of the filter paper, the membrane, the gel, 5 pieces of the filter paper, and cathode.
10. Remove air bubbles from the stack of the gel–membrane–paper by evenly compressing it.
11. Run the electroblot apparatus at 4 mA/cm^2 for 30 min.
12. During electroblotting, chemically synthesized probes containing locked nucleic acid (LNA) are labeled with ^{32}P (*see* the next step). For instance, sequences of the LNA probes that we used are as follows: aaGctAtcTgaGcaCctGtg for piR-13, ccTagGagAaaAtaCtaGac for piR-chr94, ccTcaAgcTttCatAtcTgt for gsRNA505, and acActGatTtcAaaTggTgc for mouse miR-29b-1 (positive control for periodate oxidation). Residues for LNA are indicated by capital letters.

13. Incubate 10 μL of reaction mixture consisting of 0.4 pmol/μL of an LNA probe, 1× Reaction buffer (attached in Optikinase), 0.037 MBq/μL of [γ-^{32}P] ATP, and 1 U/μL of Optikinase at 37 °C for 30 min, and then keep the mixture on ice until use.
14. After electroblotting, air-dry and UV-crosslink (254 nm, 1,200 μW/cm^2, twice) the blotted membrane.
15. Incubate the blot at 42 °C for 2 h with 10 mL of ULTRAhyb-Oligo in a hybridization bag.
16. Recover the ULTRAhyb-Oligo buffer in a 15 mL tube and transfer the blot into a new hybridization bag. To avoid drying the blot until starting hybridization, carry out **steps 17** and **18** immediately.
17. Add 5 μL of the 5′-^{32}P labeled LNA probe solution to the recovered ULTRAhyb-Oligo buffer and mix gently.
18. Pour the solution into the new hybridization bag containing the blot and seal it.
19. Incubate the blot at 42 °C for 21 h with the ^{32}P-labeling buffer in the new hybridization bag. After hybridization, discard the buffer appropriately.
20. Wash the blot with 50 mL of 2× SSC twice at 42 °C for 30 min.
21. Air-dry the blot and wrap it with a wrapping film.
22. Visualize the blot using the FLA-7000 imaging system.
23. Expected results are that piRNA species are unaffected by the periodate oxidation and β-elimination. By contrast, miRNA species treated with periodate oxidation and β-elimination migrate faster than untreated samples, because miRNA without the 3′-terminus 2′-*O*-methylation has a 2′,3′-*cis*-diol group, and it is chemically truncated at the 3′-termini by those treatments.

3.6 LC/MS Analysis of 3′-Terminal Nucleosides of piRNAs

1. The piRNA fraction (approximately 0.8 μg) was digested by RNase T_2 into 5′,3′-diphospho nucleotides, 3′-monophospho nucleotides, and 3′-terminal (2′-*O*-methyl)nucleosides (pNp, Np, and Nm, respectively, in Scheme 1a). "p" stands for phosphate group and "Nm" stands for 2′-*O*-methylated nucleoside (N = A, C, G, or U) in 20 μL of a reaction mixture containing 40 mM ammonium acetate (pH 5.3) and 1.25 units/mL RNase T_2 at 37 °C for 3 h (*see* **Notes 15** and **16**).
2. The RNA digests were subjected to the micro-flow LC/MS system at a flow rate of 0.15 mL/min. The mobile phase consisted of 5 mM NH_4OAc (pH 5.1) (solvent A) and 60 % acetonitrile (solvent B). Linear gradients were programmed as follows: 1–35 % B in 0–35 min, 35–99 % B in 35–40 min, 99 % B in 40–50 min, 99–1 % B in 50–50.1 min, and 1 % B in 50.1–60 min. The eluates were analized by a m/z range of 103–700 in positive ion detection mode.

3. Analyze the data. Typical result is shown in Fig. 1 and in published literature [5, 8, 9, 24].

3.7 LC/MS Analysis of piRNA Fragments

1. The piRNA fraction (approximately 2 μg) was digested at 37 °C for 30 min in 10 μL of a reaction mixture containing 10 mM ammonium acetate (pH 5.3) and 1 unit/μL RNase T_1, or 10 mM ammonium acetate (pH 7.7) and 1 ng/μL RNase A (*see* **Note 17**).
2. After reaction, an equal volume of 0.1 M triethylamine-acetate (TEAA) (pH 7.0) was added to the reaction mixture (*see* **Notes 16** and **18**).
3. The RNA fragments produced by RNase T_1 or RNase A digestion were subjected to the nano-flow LC/MS system of QSTAR XL and DiNa LC at a flow rate of 500 nL/min. The mobile

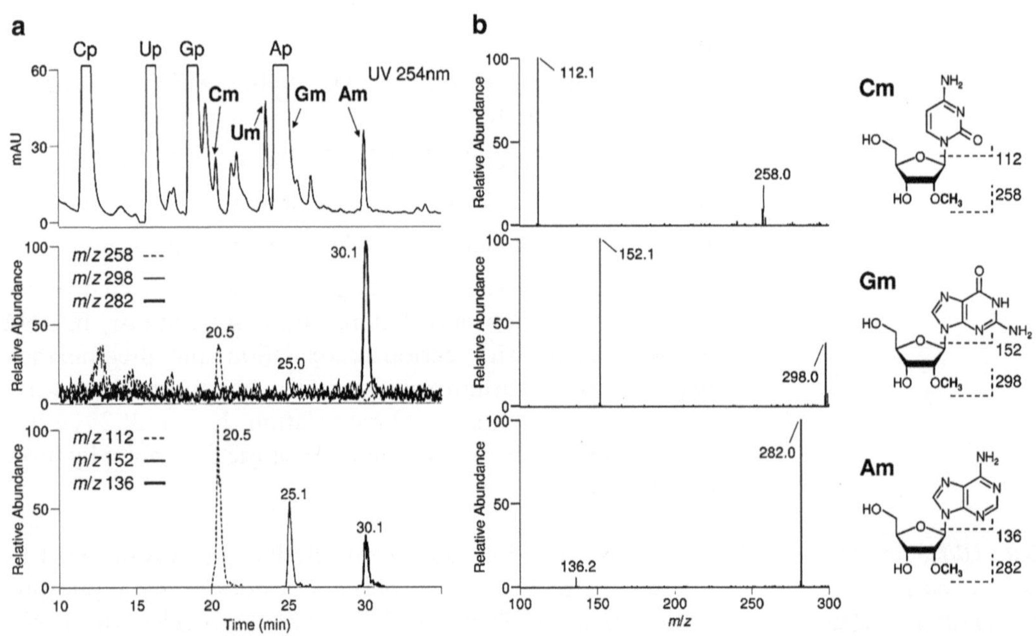

Fig. 1 Mass spectrometric analysis of piRNA digests produced by RNase T_2. (**a**) *Top panel* shows UV chromatogram at 254 nm. In addition to unmodified 3′-monophospho nucleotides (Cp, Up, Gp, and Ap), three kinds of 2′-*O*-methylnucleosides (Cm, Um, and Am) are detected. 2′-*O*-Methylguanosine (Gm), which is hidden due to overlapping of the peak of Ap, also is indicated. *Middle panel* shows the ion-extracted chromatograms for protonated molecular ions of Cm (*m/z* 258, *dashed line*), Gm (*m/z* 298, *gray line*), and Am (*m/z* 282, *black line*). *Bottom panel* shows the ion-extracted chromatograms for nucleobase-related product ions of Cm (*m/z* 112, *dashed line*), Gm (*m/z* 152, *gray line*), and Am (*m/z* 136, *black line*). In the *middle* and *bottom panels*, Um is not detectable because of the inefficiency of protonation for electrospray ionization. The retention time of each 2′-*O*-methylnucleoside also can be confirmed by comparison with commercially available standards. (**b**) Product ion spectra of 2′-*O*-methylnucleosides, Cm (*top*), Gm (*middle*), and Am (*bottom*). The chemical structure and the assignments for product ions of the 2′-*O*-methylnucleosides are depicted at the *right side* of the spectra

phase consisted of 0.4 M HFIP (pH 7.0 adjusted with triethylamine) (solvent A) and 0.4 M HFIP (pH 7.0 adjusted with triethylamine) with 50 % methanol (solvent B). Linear gradients were performed as follows: 0–80 % B in 0–40 min, 80 % B in 40–44 min, 80–0 % B in 44–46 min, and 0 % B in 46–65 min. The eluates were analized by a m/z range of 600–2000 in negative ion detection mode.

4. Analyze the data. Typical results are shown in Fig. 2.

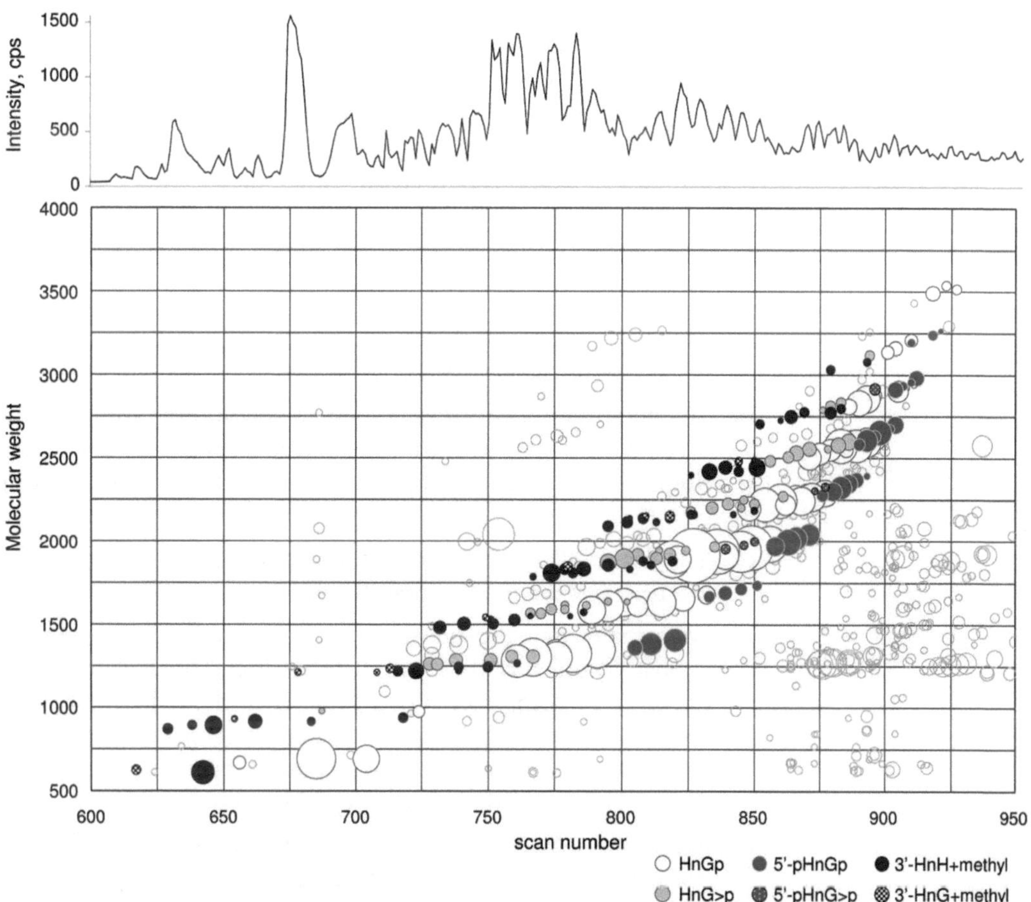

Fig. 2 Mass spectrometric analysis of piRNA fragments digested by RNase T_1. The *x*-axis shows retention time indicated as scan number. The *y*-axis shows molecular weight of detected fragments deconvoluted from *m/z* of their multivalent ions. A base peak chromatogram is shown above the 2D plot. Detected ions are indicated in each *circle*, the size of which is proportional to its intensity. According to the exact molecular mass obtained by highly accurate MS analysis, the nucleobase composition with or without modifications can be determined. The fragments are classified into three types: 5′-terminal origin (*gray* and *hatching pattern with gray*), internal origin (*white* and *light gray*), and 3′-terminal origin (*solid dark gray* and *checkerboard pattern*). The 3′-terminal fragments are found to have a single methylation, whereas the 5′-terminal and internal fragments possess no modifications. No 3′-terminal fragments without methylation are detected, suggesting that 3′-termini of piRNAs are fully modified by 2′-*O*-methylation. To determine the position that is 2′-*O*-methylated, collision-induced dissociation (CID) analysis for each fragment is required [5]. The H indicates A, C or U

4 Notes

1. Avoid thawing tissues during manipulation.
2. This step helps to decrease contaminants and to separate the interphase easily in later steps.
3. Chloroform can be used instead of CIA.
4. The upper aqueous phase is usually colored yellow.
5. Additional chloroform extractions can remove residual proteins and some kinds of hydrophobic contaminants. Temperature should be kept low.
6. For higher recovery of small RNA, instead of isopropyl alcohol, the ethanol precipitation procedure should be chosen.
7. When the pellet cannot be dissolved in a short time, incubate it on ice or at room temperature for 1 h.
8. A second chloroform extraction can remove many kinds of hydrophobic contaminants in the absence of phenol. This treatment gives superior solubility of the RNA mixture.
9. This is a clean-up step that can remove remaining polysaccharide-like contaminants.
10. To better remove phenol and guanidinium salt, ethanol precipitation may be needed again (repeat **steps 35–39** twice). **Steps 35–39** may be skipped when the RNA pellet is sufficiently small.
11. The use of 80 % ethanol is recommended instead of 70 % ethanol to avoid the loss of small RNA.
12. We usually dry the pellet under vacuum until the pellet is almost completely (not completely) dried. It takes longer to dissolve the completely dried pellet than the pellet that is not completely dried. Due to the intensive removal of contaminants, the solubility of RNA prepared by this procedure is higher than commonly used procedures.
13. Additional centrifugation is required if the RNA solution did not pass through the filter sufficiently.
14. Three to four volumes of ethanol should be added for the ethanol precipitation of small RNA (<50 nucleotides) to achieve a quantitative recovery.
15. Because RNase T_2 has been commercially unavailable, RNase I (e.g., Promega, M4261) may be used instead of RNase T_2. We have not optimized the digestion condition for RNase I yet.
16. Digestion of RNA may form 2′,3′-cyclic phosphate instead of 3′-monophosphate. The 2′,3′-cyclic phosphate form is denoted as "N>p" in Fig. 2.
17. piRNA molecules have 5′-monophosphate and 3′-hydroxyl groups in their sugar-phosphate backbone (Scheme 1). Because

a RNase T_2 digestion of piRNAs

5'- pNpN...pNpNm -3' → pNp + Np + Np + ...+ Np + Nm

b RNase T_1 digestion of piRNAs

5'- pNpN...pNpNm -3' → pHpH...pHpGp + HpH...pHpGp + ... + HpH...pHpGp + HpH...pHpNm

c RNase A digestion of piRNAs

5'- pNpN...pNpNm -3' → pRpR...pRpYp + RpR...pRpYp + ... + RpR...pRpYp + RpR...pRpNm

Scheme 1 (**a**) RNase T_2 digestion of piRNAs. (**b**) RNase T_1 digestion of piRNAs. (**c**) RNase A digestion of piRNAs

RNase T_1 or RNase A cleaves substrates into 3′-phosphate and 5′-hydroxyl forms, digestion of piRNAs produces three types of molecules derived from a 5′-terminal fragment, internal fragments, and a 3′-terminal fragment (Scheme 1b, c, H = A, C or U; R = A or G; Y = C or U).

18. The reaction mixture can be stored at −80 °C for a few weeks. TEAA is added just before LC/MS analysis.

Acknowledgments

We sincerely appreciate the Suzuki laboratory members for fruitful discussion on this project. This work was supported by Grants-in-Aid for Scientific Research on Priority Areas from the Ministry of Education, Science, Sports, and Culture of Japan (to Takeo S. and Tsutomu S.) and by a grant from the New Energy and Industrial Technology Development Organization (NEDO) (to Tsutomu S.).

References

1. Juliano C, Wang J, Lin H (2011) Uniting germline and stem cells: the function of Piwi proteins and the piRNA pathway in diverse organisms. Annu Rev Genet 45:447–469
2. Siomi MC, Sato K, Pezic D, Aravin AA (2011) PIWI-interacting small RNAs: the vanguard of genome defence. Nat Rev Mol Cell Biol 12:246–258
3. Pillai RS, Chuma S (2012) piRNAs and their involvement in male germline development in mice. Dev Growth Differ 54:78–92
4. Kirino Y, Mourelatos Z (2007) Mouse Piwi-interacting RNAs are 2′-O-methylated at their 3′ termini. Nat Struct Mol Biol 14: 347–348
5. Ohara T, Sakaguchi Y, Suzuki T, Ueda H, Miyauchi K (2007) The 3′ termini of mouse Piwi-interacting RNAs are 2′-O-methylated. Nat Struct Mol Biol 14:349–350
6. Horwich MD, Li C, Matranga C, Vagin V, Farley G, Wang P, Zamore PD (2007) The Drosophila RNA methyltransferase, DmHen1,

modifies germline piRNAs and single-stranded siRNAs in RISC. Curr Biol 17:1265–1272
7. Houwing S, Kamminga LM, Berezikov E, Cronembold D, Girard A, van den Elst H, Filippov DV, Blaser H, Raz E, Moens CB, Plasterk RH, Hannon GJ, Draper BW, Ketting RF (2007) A role for Piwi and piRNAs in germ cell maintenance and transposon silencing in Zebrafish. Cell 129:69–82
8. Zhou X, Zuo Z, Zhou F, Zhao W, Sakaguchi Y, Suzuki T, Cheng H, Zhou R (2010) Profiling sex-specific piRNAs in zebrafish. Genetics 186:1175–1185
9. Saito K, Sakaguchi Y, Suzuki T, Suzuki T, Siomi H, Siomi MC (2007) Pimet, the Drosophila homolog of HEN1, mediates 2′-O-methylation of Piwi-interacting RNAs at their 3′ ends. Genes Dev 21:1603–1608
10. Yu B, Yang Z, Li J, Minakhina S, Yang M, Padgett RW, Steward R, Chen X (2005) Methylation as a crucial step in plant microRNA biogenesis. Science 307:932–935
11. Yang Z, Ebright YW, Yu B, Chen X (2006) HEN1 recognizes 21–24 nt small RNA duplexes and deposits a methyl group onto the 2′ OH of the 3′ terminal nucleotide. Nucleic Acids Res 34:667–675
12. Huang Y, Ji L, Huang Q, Vassylyev DG, Chen X, Ma JB (2009) Structural insights into mechanisms of the small RNA methyltransferase HEN1. Nature 461:823–827
13. Zhao Y, Mo B, Chen X (2012) Mechanisms that impact microRNA stability in plants. RNA Biol 9:1218–1223
14. Huang RH (2012) Unique 2′-O-methylation by Hen1 in eukaryotic RNA interference and bacterial RNA repair. Biochemistry 51:4087–4095
15. Kirino Y, Mourelatos Z (2007) The mouse homolog of HEN1 is a potential methylase for Piwi-interacting RNAs. RNA 13:1397–1401
16. Kamminga LM, Luteijn MJ, den Broeder MJ, Redl S, Kaaij LJ, Roovers EF, Ladurner P, Berezikov E, Ketting RF (2010) Hen1 is required for oocyte development and piRNA stability in zebrafish. EMBO J 29: 3688–3700
17. Billi AC, Alessi AF, Khivansara V, Han T, Freeberg M, Mitani S, Kim JK (2012) The Caenorhabditis elegans HEN1 ortholog, HENN-1, methylates and stabilizes select subclasses of germline small RNAs. PLoS Genet 8:e1002617
18. Okamura K, Chung WJ, Ruby JG, Guo H, Bartel DP, Lai EC (2008) The Drosophila hairpin RNA pathway generates endogenous short interfering RNAs. Nature 453:803–806
19. Kurth HM, Mochizuki K (2009) 2′-O-methylation stabilizes Piwi-associated small RNAs and ensures DNA elimination in Tetrahymena. RNA 15:675–685
20. Kawaoka S, Izumi N, Katsuma S, Tomari Y (2011) 3′ end formation of PIWI-interacting RNAs in vitro. Mol Cell 43:1015–1022
21. Ameres SL, Horwich MD, Hung JH, Xu J, Ghildiyal M, Weng Z, Zamore PD (2010) Target RNA-directed trimming and tailing of small silencing RNAs. Science 328:1534–1539
22. Suzuki T, Ikeuchi Y, Noma A, Sakaguchi Y (2007) Mass spectrometric identification and characterization of RNA-modifying enzymes. Methods Enzymol 425:211–229
23. Kimura S, Ikeuchi Y, Kitahara K, Sakaguchi Y, Suzuki T, Suzuki T (2012) Base methylations in the double-stranded RNA by a fused methyltransferase bearing unwinding activity. Nucleic Acids Res 40:4071–4085
24. Suzuki T, Sakaguchi Y, Suzuki T (2007) Mass spectrometric analysis of 3′-terminal nucleosides of non-coding RNAs. Protoc Exch. doi:10.1038/nprot.2007.1185

Chapter 7

HITS-CLIP (CLIP-Seq) for Mouse Piwi Proteins

Anastassios Vourekas and Zissimos Mourelatos

Abstract

Piwi proteins, such as Aubergine in *Drosophila* and Miwi and Mili in mice, form a major subclade of the Argonaute family, which comprise a distinct class of RNA-binding proteins (RBPs) able to bind small RNAs. Small RNAs can target complementary RNAs. Piwis are essential for the animal germline and bind Piwi-interacting RNAs (piRNAs) to form pi-RiboNucleoProteins (piRNPs). Although many piRNAs target retrotransposons for safeguarding genome integrity of the germ cell, whether piRNAs can target other mRNAs for regulatory purposes is still under investigation. Here we present the technical protocol for "High Throughput Sequencing after in vivo Crosslinking and Immunoprecipitation" (HITS-CLIP, CLIP-Seq), adapted for mouse Piwi proteins Mili and Miwi. We also provide general recommendations for the application of this protocol for different RBPs and also for the bioinformatic analysis of the deep sequencing data.

Key words piRNA, piRNP, Argonaute, Piwi, Mili, Miwi, Miwi2, Aub, Ago3, Next gen sequencing, Illumina, cDNA, Immunoprecipitation, HITS-CLIP, CLIP-Seq, RNA-IP, T4 RNA ligase, Reverse transcriptase, Polymerase chain reaction, PCR, RT-PCR, Post-transcriptional RNA processing, Gene silencing

1 Introduction

Germ cells employ complex mechanisms to process, transport and localize, and regulate stability and translation of mRNAs that control their developmental program. Piwi proteins are essential for animal germline development and they bind piRNAs to silence retrotransposons in the germline [1, 2]. However, significant numbers of piRNAs are unique [3–5] and it is unknown whether piRNAs can use seed sequence complementarity to target RNAs other than repeat elements. *Mus musculus* Mili [6], Miwi [7], Tdrd's [8–11], and MVH [12, 13], which are expressed in postnatal testis, localize in RNA-rich, dense cytoplasmic foci that are collectively called germ granules—nuage [14]. The germ granules have been implicated in translational control and mRNA stability, but the underlying mechanisms are not understood. HITS-CLIP is the method of choice for the identification of the identities of the

Mikiko C. Siomi (ed.), *PIWI-Interacting RNAs: Methods and Protocols*, Methods in Molecular Biology, vol. 1093, DOI 10.1007/978-1-62703-694-8_7, © Springer Science+Business Media, LLC 2014

RNAs targeted by RNA-binding proteins (RBPs), but also the exact nucleotide sequences, at unprecedented resolution [15–17]. We recently undertook Mili and Miwi HITS-CLIP and biochemical characterization of testis RNPs [18], thus—(a) providing the first in vivo snapshots of piRNA precursor processing that lead to *a model for piRNA biogenesis*; (b) revealing a *piRNA independent formation of Miwi mRNPs* critical for spermiogenesis. Here, we describe the adaptation of the HITS-CLIP protocol [15], for the identification of the in vivo RNA bound by mouse Piwi proteins. General considerations for the use of this protocol with Piwi or other RBPs from various sources are suggested throughout the text, and also some guidelines for the bioinformatic analysis of the next generation sequencing data.

2 Materials

2.1 Tissue Harvesting and UV Crosslinking

1. Freshly harvested mouse testes.
2. Ice-cold HBSS (Life Technologies).
3. 1× PBS (no Mg^{2+}/Ca^{2+}) (Roche).
4. Stratalinker (Model 1800 or 2400; Stratagene).

2.2 Preparation of Cell Lysate, Immunoprecipitation, and Labeling of Crosslinked Protein–RNA Complexes

1. Dynabeads® protein A (Life Technologies).
2. 1 M Na-phosphate buffer (pH 8.0): Mix 1 volume of 1 M NaH_2PO_4 with 13.7 volumes of 1 M Na_2HPO_4; adjust to pH 8.0 by adding NaH_2PO_4 for increasing acidity, or Na_2HPO_4 for increasing alkalinity.
3. Antibody (Ab) binding buffer: 0.1 M Na-phosphate (pH 8.0), 0.1 % IGEPAL CA-630, 5 % Glycerol.
4. Rabbit anti-mouse IgG Fcγ Fragment Specific (Jackson ImmnoResearch).
5. Nonimmune mouse serum or mouse IgG.
6. Anti-Piwi protein antibodies. We have used our own anti-Mili (17–8 mouse monoclonal antibody) and anti-Miwi (rabbit polyclonal) and Cell signaling Technologies G82 anti-Miwi antibody with success.
7. 1× PMPG buffer: 1× PBS, 2 % Empigen (Sigma).
8. 5× PMPG wash buffer: 5× PBS, 2 % Empigen.
9. Lysis buffer (prepare fresh each time): 1× PMPG plus one tablet of Complete Mini EDTA-free Protease Inhibitor Cocktail Tablets (Roche) per 10 mL and 1 U/μL rRNasin (Promega).
10. RQ1 DNase (Promega).
11. RNase T1 (Roche).
12. Polyallomer Microfuge tubes (for tabletop ultracentrifuge, Beckman, 357448).

13. 4× SDS reducing loading buffer: 1 mL of NUPAGE SDS loading buffer (Invitrogen) supplemented with 100 μL of β-mercaptoethanol (Bio-Rad).
14. MilliQ RNase-free water.
15. ^{32}P-γ-ATP (10 μCi/μL, 3,000 Ci/mmol).
16. 1× PNK buffer: 50 mM Tris–HCl (pH 7.4), 10 mM $MgCl_2$, 0.5 % IGEPAL CA-630.
17. T4 Polynucleotide Kinase (T4 PNK; NEB).
18. Illustra Microspin G-25 columns (GE Healthcare).
19. Beckman Optima TL Ultracentrifuge with TLA 100.3 rotor.
20. Thermomixer (Eppendorf).
21. 10 mM ATP.
22. RL3(−P) RNA oligo: from IDT, at 250 nmol synthesis scale, and purified by RNase-free HPLC. The sequence is 5′-OH GUGUCAGUCACUUCCAGCGG 3′-Inverted dT. Inverted dT is blocking the 3′ end of RL3 adapters, preventing concatamerization and aberrant ligation of the RL3 adapter with extracted RNAs.

2.3 Ligation of the Labeled RL3 RNA Adapter (On-Beads)

1. MilliQ RNase-free water.
2. Antarctic Phosphatase (NEB).
3. 10× Antarctic Phosphatase buffer (NEB).
4. rRNasin (Promega).
5. 1× PNK+EGTA buffer: 50 mM Tris–HCl (pH 7.4), 20 mM EGTA, 0.5 % IGEPAL CA-630.
6. 0.2 μg/μL BSA.
7. T4 RNA Ligase (NEB).
8. 10× T4 RNA Ligase buffer (NEB).
9. 10 mM ATP.
10. 1× PMPG buffer: 1× PBS, 2 % Empigen (Sigma).
11. 5× PMPG wash buffer: 5× PBS, 2 % Empigen.
12. RL3(+P) RNA oligo: from IDT, at 250 nmol synthesis scale, and purified by RNase-free HPLC. The sequence is RL3(+P): 5′-p GUGUCAGUCACUUCCAGCGG 3′-Inverted dT. Inverted dT is blocking the 3′ end of RL3 adapters, preventing concatamerization and aberrant ligation of the RL3 adapter with extracted RNAs.

2.4 SDS-PAGE and Nitrocellulose Transfer Analysis of Crosslinked RNA–Protein Complexes

1. 1× PNK buffer: 50 mM Tris–HCl (pH 7.4), 10 mM $MgCl_2$, 0.5 % IGEPAL CA-630.
2. 4× SDS reducing loading buffer: 1 mL of NUPAGE SDS loading buffer (Invitrogen) supplemented with 100 μL of β-mercaptoethanol (Bio-Rad).

3. NuPAGE 4–12 % Bis-Tris precast gel (Invitrogen).
4. NuPAGE MOPS SDS Running Buffer (Invitrogen).
5. 10× Western Blot buffer (for semi-dry transfer): 0.625 M Tris base, 0.18 M Glycine.
6. 1× Western Blot buffer: 10 % (v/v) 10× buffer, 20 % (v/v) Methanol, 70 % (v/v) MilliQ H_2O.
7. SE 400 Sturdier Gel electrophoresis apparatus with 18×24 cm glass plates (Amersham).
8. Nitrocellulose Membrane (Invitrogen LC2001, 0.45 μm pore size).
9. Semi-dry transfer apparatus, TE-77 (Hoefer).
10. Autoradiography films: Kodak BioMax MR, 8×10 in.

2.5 RNA Extraction

1. 25 mg/mL Proteinase K (PK) solution (Roche).
2. 5× PK Buffer: 500 mM Tris–Cl pH 7.5, 250 mM NaCl, 50 mM EDTA.
3. 1× PK Buffer/7 M urea solution (prepare fresh each time).
4. Acid RNA phenol/$CHCl_3$ (Ambion).
5. Acid Phenol/$CHCl_3$/Isoamyl alcohol 25:24:1 (Sigma).
6. Chloroform:Isoamyl alcohol 24:1 (Sigma).
7. 5 mg/mL Glycogen (Ambion).
8. 3 M Sodium Acetate (pH 5.2) (Ambion).
9. 1:1 (v/v) Ethanol/isopropanol solution.

2.6 5′ Adapter (RL3) Ligation

1. 70 % Ethanol.
2. T4 RNA Ligase (Fermentas).
3. 10× T4 RNA Ligase buffer (Fermentas).
4. rRNasin (Promega).
5. 3 M Sodium Acetate (pH 5.2) (Ambion).
6. 0.2 μg/μL BSA.
7. RQ1 DNase.
8. 10 mM ATP.
9. Acid Phenol/chloroform/isoamyl alcohol (25:24:1) (Sigma).
10. 1:1 (v/v) Ethanol/isopropanol solution.
11. RL5 RNA oligo: from IDT, at 250 nmol synthesis scale, and purified by RNase-free HPLC. The sequence is 5′-OH AGGGAGGACGAUGCGG 3′-OH.

2.7 Denaturing PAGE/7 M Urea Electrophoretic Analysis of Extracted RNA

1. 10 % Urea/PAGE solution (1 L): combine in a glass beaker: 480 g Urea, 250 mL of 40 % Acrylamide/Bis (19:1), 100 mL 10× TBE (Ambion), and water up to 1 L. Stir until completely dissolved, filter-sterilize, and store up to a year at RT in an aluminum foil-covered bottle (to protect from light).
2. 20 % Urea/PAGE (1 L): combine in a glass beaker: 480 g Urea, 500 mL of 40 % Acrylamide/Bis (19:1), 100 mL 10× TBE, and water up to 1 L. Stir until completely dissolved, filter-sterilize and store as described above.

2.8 cDNA Synthesis and Preparation of Deep Sequencing Libraries

1. dNTP mix.
2. 0.1 M DTT.
3. Superscript III Reverse Trascriptase (RT; Invitrogen).
4. Accuprime Pfx Supermix (Invitrogen).
5. MetaPhor Agarose (Lonza).
6. 1× TAE buffer.
7. Ethidium bromide.
8. QIAquick Gel Extraction Kit (Qiagen).
9. DNA primers: from IDT, at 250 nmol synthesis scale, and PAGE purified.

   DP5: 5′-AGGGAGGACGATGCGG-3′.

   DP3: 5′-CCGCTGGAAGTGACTGACAC-3′.

   DSFP5: 5′-AATGATACGGCGACCACCGACTATGGATAC TTAGTCAGGGAGGACGATGCGG-3′.

   DSFP3: 5′-CAAGCAGAAGACGGCATACGACCGCTGGA AGTGACTGACAC-3′.

   SSP1: 5′-CTATGGATACTTAGTCAGGGAGGACGATGCGG-3′.

3 Methods

The outline of the procedures and representative figures from experiments are shown in Fig. 1. All procedures and centrifugations are performed on ice or at 4 °C unless otherwise indicated. Use RNase-free solutions, tubes, and pipettes.

3.1 Tissue Harvesting and UV Crosslinking (Day 1)

1. Harvest testes in batches of four and keep them covered in ice-cold HBSS until harvest is complete. Remove tunica albuginea using finely tipped forceps, and transfer testes into a glass dounce homogenizer. Triturate tissue by mild mechanical disruption (i.e., using a smaller size pestle and pipetting), with care not to lyse too many cells (*see* **Note 1**).

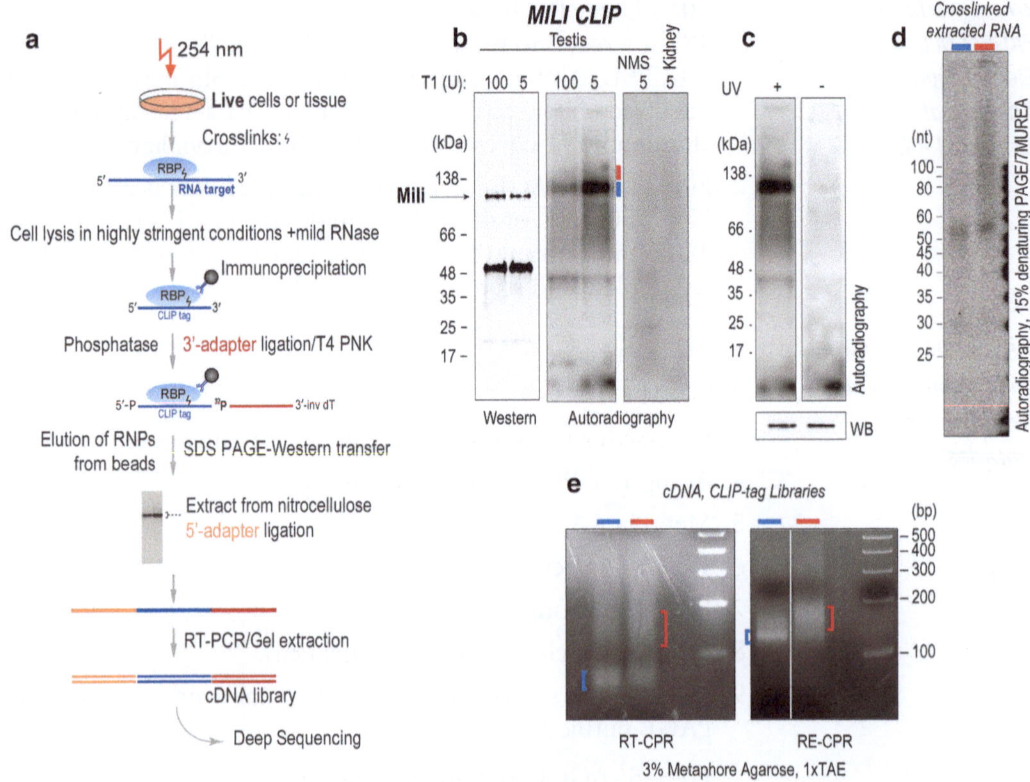

Fig. 1 Mili HITS-CLIP. (**a**) Graphic overview of the HITS-CLIP experimental protocol. (**b**) Autoradiogram of ^{32}P labeled Mili protein–RNA complexes after in vivo UV-crosslinking, immunoprecipitation, labeling of the crosslinked RNA, and SDS-PAGE analysis. Western blot analysis of the same samples using the same anti-Mili antibody is also shown for comparing the difference of electrophoretic mobility between the bulk of the protein, which is not crosslinked (identified by western blot), and the crosslinked Protein–RNA complexes in the autoradiogram. CLIP using lysates subjected to mild RNase T1 treatment (5 U) and moderate RNase T1 treatment (100 U) is shown. Note the increased radioactive signal in mildly treated sample compared to moderately treated. RNA (CLIP tags) were extracted from the membranes after cutting at indicated areas: a *blue line* marks the major radioactive signal containing mainly protein–piRNA complexes; a *red line* marks larger Protein–RNA complexes which appear as a smear extending to higher molecular weights. Negative control CLIPs were performed using nonimmune mouse serum (NMS) with crosslinked testis, and anti-Mili antibodies with crosslinked kidney. (**b**) Comparison of UV-crosslinked and non-crosslinked labeled Mili Protein–RNA complexes. Absence of UV-crosslinks results in loss of the specific radioactive signal that is migrating slightly slower than the Mili western blot signal. (**c**) Denaturing PAGE electrophoretic analysis of Mili crosslinked and extracted RNA. Observe enrichment in piRNAs for the sample extracted from the main radioactive signal (*blue line*), and enrichment in large RNAs for the sample extracted from the slower migrating complexes (*red line*). (**d**) RT-PCR and RE-PCR amplification of Mili crosslinked and extracted RNA. PCR products were gel purified after Metaphor agarose gel electrophoresis. *Blue* and *red brackets* denote piRNA and large tag enriched cDNA samples, respectively

2. Transfer tissue suspension in petri dish and keep it be covered with ice-cold HBSS (200 mg of tissue—four adult testies—in a 10 cm plate, 8 mL of HBSS). Irradiate tissue suspension three times at 400 mJ/cm^2 in Stratalinker, with 30-s intervals for cooling. Mix suspension between each irradiation.

Always keep on ice. An important negative control is tissue that has not been exposed to UV. Other negative controls include tissues that do not normally express the target protein (in the case of Piwi any somatic tissue) and also testes from Piwi knockout mice.

3. Collect cell suspension in a 15-mL falcon tube and pellet at 1200 × *g* for 10 min at 4 °C. Resuspend the cell pellet in 1 mL of PBS and divide it in two eppendorf tubes (each aliquot is good for one CLIP). Spin at 1,000 × *g* for 30 s at 4 °C, remove supernatant, freeze pellets in liquid nitrogen and keep at −80 °C until use (*see* **Notes 2** and **3**).

3.2 Preparation of Cell Lysate, Immunoprecipitation, and Labeling of Crosslinked Protein–RNA Complexes (Day 2)

3.2.1 Preparation of Antibody-Bound Beads

For bead washing steps in this and subsequent sections, use 1 mL of buffer. For each IP use 150 μL of protein A Dynabeads slurry. Normal (pre-immune) rabbit or mouse serum can be used in a negative control IP.

1. Wash beads three times with Ab binding buffer.
2. Resuspend beads in 350 μL Ab binding buffer.
   (a) For using a *mouse monoclonal anti-Piwi Ab* for immunoprecipitation, add 3 μL rabbit anti-mouse (bridging) Ab to the beads.
      - Rotate tubes at room temperature for 45 min.
      - Wash three times with Ab binding buffer.
      - Resuspend beads in 350 μL Ab binding buffer and add 3–5 μL of ascites (or pre-immune serum) fluid or 5–10 μg of Ab in solution.
      - Rotate tubes at 4 °C for 3 h.
      - Wash one time with Ab binding buffer and two times with 1× PMPG buffer; if you are not yet ready to add crosslinked lysate, leave beads in last washing buffer.

   (b) For using a *rabbit polyclonal anti-Piwi Ab* for immunoprecipitation, add 10 μg of Ab to the beads (or 3–5 μL of pre-immune rabbit serum).
      - Rotate tubes at 4 °C for 2–3 h.
      - Wash one time with Ab binding buffer and two times with 1× PMPG buffer.

3.2.2 Preparation of Cell Lysate

1. Resuspend each pellet of UV treated tissue using 350 μL of Lysis buffer (*see* **Note 4**). With a 1 mL pipettor, mix until flow is unforced, with care not to foam. Let sit on ice for 10 min (*see* **Note 5**).
2. Add 10 μL of RQ1 DNase to each tube and incubate at 37 °C for 5 min on a Thermomixer at 1,000 rpm. Thoroughly mix lysates by pipetting two times.

3. Prepare dilutions of RNase T1 in 1× PMPG. The lysate is subjected to RNase treatment so that the size of long crosslinked RNAs is reduced to 50–150 nucleotides (*see* **Note 6**). The conditions of RNase treatment are critical for success and should be optimized for every different experimental setting (tissue and protein expression levels): overdigested RNA–protein complexes may not be labeled efficiently in subsequent steps (*see* **Note 7**), while undigested complexes may be too big to analyze in subsequent steps. A series of high to low RNase treatment conditions should be tested, to identify the conditions in which crosslinked RNPs give a strong and specific signal after labeling and SDS-PAGE analysis. For high RNase treatment use 1 μL of undiluted RNase T1 per 350 μL of lysate. Useful dilutions include 1/10, 1/20, 1/100, 1/1,000, and also no RNase. For mouse testis and Piwi protein CLIP, crosslinked RNAs are sufficiently cleaved by the activities of endogenous nucleases, despite the presence of rRNasin and handling at 4 °C or on ice, and therefore no exogenous RNase is required for this step.
4. For RNase treatment, add 1 μL of 1× PMPG or RNase T1 dilution per 350 μL of lysate and incubate at 37 °C for 5 min on a Thermomixer at 1,000 rpm. Thoroughly mix lysates by pipetting two times.
5. Transfer lysates to polyallomer Microspin tubes for ultracentrifugation and weigh on a precision scale, for accurate balancing. Spin lysates in prechilled ultracentrifuge at 90,000 × *g* for 30 min at 4 °C (*see* **Note 8**).
6. Carefully remove the supernatant (S90) and keep on ice. As a pre-immunoprecipitation sample save 15 μL of the S90; mix with equal volume (15 μL) of 4× SDS reducing loading buffer and incubate at 70 °C for 12 min.

3.2.3 Immunoprecipitation

1. Add the S90 lysate to a tube containing Ab-bound Dynabeads from in Subheading 3.2.1 (after removing last wash buffer).
2. Rotate beads with lysate for 3 h at 4 °C (During this step, you may carry on Subheading 3.2.4).
3. Remove the supernatant (Piwi protein depleted S90, keep at −80 °C) and save 15 μL for immunoblot analysis.
4. Wash beads with ice-cold buffer:
   (a) Two times with 1× PMPG (Wash Buffer)
   (b) Two times with 5× PMPG (High-salt Wash Buffer)
   (c) Two times with 1× PNK Buffer

3.2.4 ^{32}P Labeling of the 3′ RNA Adapter (RL3)

You may perform this step during the immunoprecipitation (Subheading 3.2.3) or the previous day. Use fresh ^{32}P-γ-ATP.

The reaction described below is for preparing enough labeled RL3 adapter for ten subsequent ligations to immunoprecipitated piRNPs.

1. Combine the following (total reaction volume 50 μL):
   (a) 6.5 μL MilliQ RNase-free H_2O
   (b) 2.5 μL 50 pmol/μL RL3(–P) linker
   (c) 3 μL rRNasin
   (d) 25 μL ^{32}P-γ-ATP (10 μCi/μL, 3,000 Ci/mmol)
   (e) 5 μL 10× PNK Buffer
   (f) 8 μL T4 PNK enzyme
2. Incubate at 37 °C for 30 min.
3. Add 2 μL of 1 mM ATP, and incubate at 37 °C for 5 more min (*see* **Note 9**). Spin down and place on ice.
4. Resuspend the resin in the G-25 column by inverting several times. Pre-spin the column for 1 min at 735 ×*g*, apply the sample to resin, and spin the column for 2 min at 735 ×*g*. Collect the eluate (labeled RL3 adapter) in a clean tube. Probe RL3 eluate and Mini spin-column with a Geiger counter. Typically, RL3 solution accounts for more than 60–70 % of total radioactive signal.
5. Heat inactivate the residual T4 PNK in the eluted labeled RL3 at 65 °C for 20 min (*see* **Note 10**). If you are not ready to use, store RL3 adapter at −20 °C.

3.3 Ligation of the Labeled RL3 RNA Adapter (On-Bead)

3.3.1 Phosphatase Treatment of the Crosslinked RNAs (On-Bead)

1. Prepare the following reaction (one per CLIP sample, total reaction volume 80 μL).
   (a) 67 μL MilliQ RNase-free H_2O
   (b) 8 μL 10× Antarctic phosphatase buffer
   (c) 3 μL Antarctic phosphatase
   (d) 2 μL rRNasin
2. Add 80 μL of phosphatase reaction mixture to each tube of beads from **step 4** in Subheading 3.2.3, after removing the last washing buffer.
3. Incubate at 37 °C for 20 min on a Thermomixer at 1,000 rpm (*see* **Note 11**).
4. Wash beads with ice-cold buffer:
   (a) One time with 1× PNK Buffer
   (b) One time 1× PNK+EGTA Buffer
   (c) Two times 1× PNK Buffer

3.3.2 3′ RNA Linker Ligation (On-Bead)

1. Prepare the following reaction (one per CLIP sample, total reaction volume 80 μL)
   (a) 47 μL water
   (b) 8 μL 10× T4 RNA ligase buffer
   (c) 8 μL 0.2 μg/μL BSA
   (d) 8 μL 10 mM ATP
   (e) 2 μL rRNasin
   (f) 2 μL T4 RNA ligase
   (g) 5 μL ^{32}P labeled RL3 (prepared in Subheading 3.2.4)
2. Add 80 μL of ligase reaction mix to each tube of beads from previous step, after removing the washing buffer.
3. Incubate at 16 °C for 1 h on a Thermomixer at 1,000 rpm.
4. Add 4 μL of RL3(+P) (20 pmol/μL), (*see* **Note 12**), and proceed with overnight incubation at 16 °C on a Thermomixer at 1,000 rpm.

(Day 3)

5. Wash beads with ice-cold buffer:
   (a) One time with 1× PMPG (Wash Buffer)
   (b) One time with 5× PMPG (High-salt Wash Buffer)
   (c) Two times with 1× PNK Buffer

3.3.3 T4 PNK Treatment of Crosslinked RNAs (On-Bead)

This treatment restores a 5′ phosphate end of the crosslinked RNAs, which is required for ligation of the 5′ RNA adapter (Subheading 3.6).

1. Combine the following reagents for the phosphorylation reaction (80 μL total):
   (a) 65 μL water
   (b) 8 μL 10× PNK Buffer
   (c) 2 μL rRNasin
   (d) 1 μL 10 mM ATP
   (e) 4 μL T4 PNK enzyme
2. Add 80 μL of PNK mix to each tube with beads (after removing last wash) and incubate at 37 °C for 20 min on a Thermomixer at 1,000 rpm.
3. Wash beads with ice-cold buffer:
   (a) One time with 1× PMPG (Wash Buffer)
   (b) One time with 5× PMPG (High-salt Wash Buffer)
   (c) Three times with 1× PNK buffer

3.4 SDS-PAGE and Nitrocellulose Transfer Analysis of Crosslinked RNA–Protein Complexes

1. After removing the final wash buffer, add to the beads 15 μL of 1× PNK and 15 μL of 4× SDS reducing loading buffer (*see* **Note 13**). Resuspend by pipetting and mild vortexing, and incubate at 70 °C for 12 min on a Thermomixer at 1,000 rpm.
2. Place tubes on a magnet and separate protein eluate from beads (*see* **Note 14**). Load up to 25 μL of sample per well of a 10 well Novex NuPAGE 4–12 % Bis-Tris gel (*see* **Note 15**). Keep 5 μL of each eluate for Western blot analysis. Include pre-stained markers in the run. Run the gel using MOPS running buffer at 170 V in the cold room.
3. Stop the run when dyes reach the lower opening of the gel cassette (for consistency between different experiments, stop gel running at the same exact point each time).
4. After gel run, open the gel cassette and transfer the gel into a container with 1× western blot buffer. Incubate for 1 min.
5. Set up Western transfer to nitrocellulose using a semi-dry apparatus (*see* **Note 16**) at constant 90 mA (per gel) for 1 h 10 min.
6. After transfer, rinse the nitrocellulose membrane in MilliQ RNase-free water, and gently blot the back side of the membrane on Kimwipes (do not dry completely).
7. Heat-seal the membrane and the gel separately in hybridization plastic bags. Heat-seal one corner of the membrane to prevent it from moving inside the bag. Attach small pieces of fluorescent sticker on the bags or use fluorescent marker (the stickers/markings will be aligned with the respective signal on the film after development).
8. Expose to film (Kodak BioMax MR) (*see* **Note 17**). A single prominent band at the MW of the immunoprecipitated protein is the ideal result. A band that gives a clear signal after 1–2 h is ideal for subsequent RNA extraction step. Longer exposure (overnight) may be required for fainter signals, but in that case combine two or three samples in one to extract sufficient RNA for subsequent steps. To monitor the efficiency of RNP transfer, compare the radioactive signal on the membrane with the remaining signal on the gel. It is expected that more than 60–70 % of the total radioactive signal is transferred on the membrane.
9. On a light table, tape the film first and lay the bag with the membrane above the film. Align the markings of the fluorescent sticker pieces with their signals on the film. Correct alignment is essential (*see* **Note 18**). You should be able to see the radioactive signal on the film, through the membrane. Tape the bag in place. Open a hole in the bag without moving the membrane.

10. Two different strategies can be followed for isolating crosslinked RNAs on Piwi proteins (or Argonautes in general). (a) To isolate small RNA population and larger RNA population separately or (b) both populations in one sample. In the first strategy (followed in this protocol), for isolation of small RNAs (piRNAs), cut (using a single use surgical blade) the part of the membrane containing the main radioactive signal (Fig. 1b, marked by a blue line).
11. Put the pieces in a clean tube (cut in small pieces ~4 mm^2 each to facilitate RNA extraction). This sample may also contain small fragments of larger RNAs. For isolation of larger RNAs, cut a part of the membrane of equal size with the previous, above the main radioactive signal (Fig. 1b, marked by a red line) (*see* **Note 19**). This sample will also contain piRNAs (*see* **Note 20**). After analysis of the pre and post IP lysates and also CLIP eluate samples by regular Western blot, and precise alignment of pre-stained markers on western blot and on the autoradiography, note that the main radioactive band (formed by small RNAs crosslinked on the protein) migrates ≥7 kDa above the apparent molecular weight of the immunoprecipitated protein (Fig. 1b) (*see* **Note 21**).

3.5 RNA Extraction

1. Prepare a proteinase K solution at 4 mg/mL (the stock solution by Roche is at 25 mg/mL), in 1× PK Buffer, and pre-incubate this solution at 37 °C for 20 min to degrade any present nucleases.
2. Add 200 μL of proteinase K solution to each tube containing membrane fragments from previous step. Incubate at 37 °C for 20 min on a Thermomixer at 1,000 rpm.
3. Add 200 μL of 1× PK/7 M Urea solution to each tube, and incubate for 20 min at 37 °C on a Thermomixer at 1,000 rpm.
4. Add 400 μL of acid RNA phenol/$CHCl_3$, and incubate for 20 min at 37 °C on a Thermomixer at 1,000 rpm.
5. Spin tubes in a tabletop centrifuge at maximum speed (~16,000 × *g*) for 5 min at room temperature. Collect aqueous (upper) phase (contains RNA).
6. Add 400 μL of $CHCl_3$/isoamyl alcohol and vortex. Centrifuge as in previous step, collect again the aqueous (upper) phase, and add the following:
   (a) 50 μL 3 M Sodium Acetate (pH 5.2)
   (b) 0.75 μL 5 mg/mL Glycogen
   (c) 1 mL 1:1 (v/v) Ethanol/Isopropanol solution
7. Mix well and precipitate overnight at −80 °C.

3.6 5′ RNA Adapter (RL5) Ligation (Day 4)

1. Spin tubes in a tabletop centrifuge at maximum speed for 30 min at 4 °C to pellet the RNA (*see* **Note 22**). The pellet should be detectably radioactive, although this is not a measure of the success of the RNA extraction step for faint radioactive signals.
2. Wash pellet with 1 mL 70 % ethanol, and centrifuge again at maximum speed for 15 min at 4 °C to recover the pellet.
3. Remove ethanol and air-dry the pellet (*see* **Note 23**).
4. Resuspend in 6.5 μL MilliQ RNase-free H_2O. Keep 1 μL for denaturing PAGE analysis.
5. Set up the following 5′ RNA adapter (RL5) ligation:
   (a) 5.5 μL resuspended RNA
   (b) 1 μL 10× T4 RNA ligase buffer
   (c) 0.5 μL rRNasin
   (d) 1 μL 0.2 μg/μL BSA
   (e) 1 μL 10 mM ATP
   (f) 0.5 μL T4 RNA ligase
   (g) 0.5 μL 40 pmol/μL RL5 RNA adapter
6. Incubate at 16 °C for 6 h.
7. Add to the reaction:
   (a) 79 μL MilliQ RNase-free H_2O
   (b) 11 μL 10× DNAse I Buffer
   (c) 5 μL rRNasin
   (d) 5 μL RQ1 DNAse
8. Incubate 37 °C for 20 min.
9. Add to the above mix:
   (a) 300 μL MilliQ RNase-free H_2O
   (b) 400 μL Acid Phenol/$CHCl_3$/Isoamyl alcohol 25:24:1
10. Vortex for 1 min. Spin tubes in a tabletop centrifuge at maximum speed (~16,000 × *g*) for 5 min at room temperature. Collect upper phase (should be approximately 400 μL).
11. Add 400 μL of $CHCl_3$/Isoamyl alcohol 24:1. Vortex, spin, and collect aqueous phase as in previous step.
12. Precipitate by adding:
    (a) 50 μL 3 M Sodium Acetate (pH 5.2)
    (b) 0.5 μL 5 mg/mL Glycogen
    (c) 1 mL 1:1 EtOH/isopropanol
13. Precipitate overnight at −80 °C.

3.7 Denaturing PAGE/7 M Urea Electrophoretic Analysis of Extracted RNA

1. Analyze 1 μL of RNA in 15 % Urea/PAGE [19]. Use radiolabeled small ssRNA 20–100 nts as marker.
2. Expose for 2–10 days (or even longer), depending on the amount of radioactivity of extracted RNAs. Observe enrichment in piRNAs (Fig. 1d, lane marked with blue) for the sample extracted from the main radioactive signal (Fig. 1b, blue line), and enrichment in larger RNAs (target mRNAs and piRNA precursor fragments; Fig. 1d, lane marked with red) in the sample extracted from higher molecular weights (Fig. 1b, red line).

3.8 cDNA Synthesis and Preparation of Deep Sequencing Libraries (Day 5)

Pellet the RL5 ligated RNA by centrifugation for 30 min at maximum speed and at 4 °C. Carefully aspirate the supernatant and wash the pellet with ice-cold 70 % ethanol. Pellet the RNA by centrifugation as before, remove the supernatant and air-dry. Resuspend the pellet in 10 μL of MilliQ RNase-free H_2O. The RNA now has 5′ and 3′ RNA adapters and can be reverse transcribed and amplified by PCR for the preparation of the deep sequencing library.

3.8.1 Reverse Transcriptase Reaction

1. Set up the following mixture:
   (a) 10 μL ligated RNA.
   (b) 2 μL 5 pmol/μL DP3 primer.
   (c) 1 μL 10 mM dNTPs.
2. Incubate at 65 °C for 5 min. Place on ice and quick spin.
3. Add to the above:
   (a) 1 μL DTT, 0.1 M
   (b) 4 μL 5× SuperScript RT Buffer
   (c) 1 μL rRNasin
   (d) 1 μL SuperScript III
4. Incubate at 50 °C for 45 min, 55 °C for 15 min, 90 °C for 5 min, leave at 4 °C.

3.8.2 Polymerase Chain Reaction

1. Set up the following PCR:
   (a) 27 μL Accuprime Pfx Supermix
   (b) 0.75 μL DP5 primer, 20 pmol/μL
   (c) 0.75 μL DP3 primer, 20 pmol/μL
   (d) 3 μL of the RT reaction
2. Perform the following PCR program:
   (a) 95°C for 2 min
   (b) 25–28 cycles of:
      - 95 °C for 20 s
      - 58 °C for 30 s
      - 68 °C for 30 s

(c) 68 °C for 5 min

(d) 4 °C hold

Perform each reaction in triplicate and combine replicate reactions before electrophoresis.

3. Speed-vac pooled samples to concentrate, and run each replicate pool on one lane on a 3 % Metaphor 1×TAE/EtBr gel at 100 V (*see* **Note 24**). Small RNAs give rise to a 60–70 bp band (piRNAs are 25–30 nucleotides plus 21 nucleotides from RL3, and 16 nucleotides from RL5) (Fig. 1e). Larger RNAs produce a smear extending from the small RNA band and upwards, usually with a more prominent population at sizes 100–200 bp (Fig. 1e) (*see* **Note 25**).
4. Cut out desired PCR products with a clean blade and transfer to a clean tube. Extract DNA with QIAquick Gel Extraction Kit (*see* **Note 26**), using 40 μL of elution buffer.

3.8.3 Re-PCR with Solexa Fusion Primers

1. Set up the following PCR reaction:

(a) 27 μL Accuprime Pfx Supermix

(b) 0.5 μL 20 pmol/μL DSFP5 primer

(c) 0.5 μL 20 pmol/μL DSFP3 primer

(d) 3 μL of gel extracted 1st PCR product.

2. Perform the following PCR program:

(a) 95 °C for 2 min

(b) 6–10 cycles of:

- 95 °C for 20 s
- 58 °C for 30 s
- 68 °C for 30 s

(c) 68 °C for 5 min

(d) 4 °C hold

Again, you can perform three replicate reactions per sample and pool them before electrophoresis. Speed-vac to concentrate, and run in a single lane on 3 % Metaphor 1×TAE/EtBr gel. DSFP primers add 97 bp to the size of the CLIP tag, so piRNAs will form a band at ~130 bp, and larger RNAs a diffuse smear up to 250 bp (Fig. 1e). Cut out desired sizes, and extract DNA with QIAquick Gel Extraction kit as in previous step.

3.9 Illumina Next Generation Sequencing and Considerations for Bioinformatic Analysis of Sequencing Data

1. Proceed with next generation sequencing with Illumina analyzer as per the manufacturer's instructions (*see* **Note 27**).
2. The first base of every CLIP tag sequenced corresponds to the first nucleotide of the crosslinked RNA molecule. The 3′ adapter sequence has to be "trimmed" off the CLIP tag sequence before alignment on the reference genome (*see* **Note 28**).

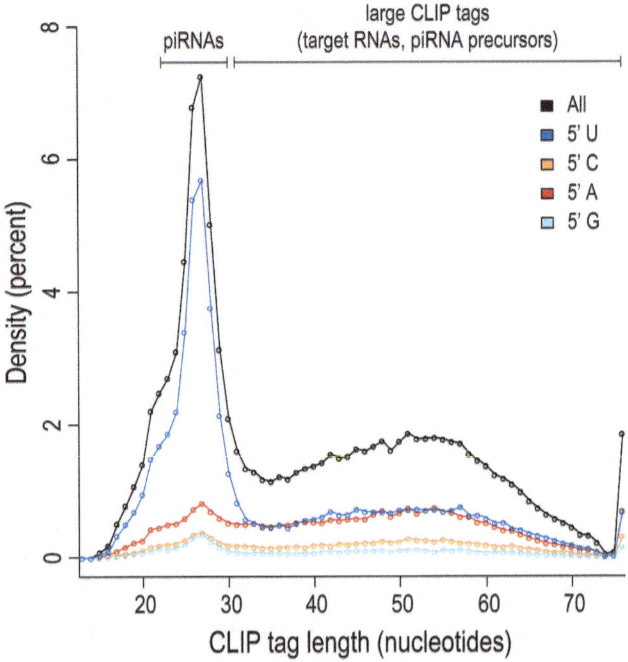

Fig. 2 Size distribution and 5′ nucleotide preference of a Mili CLIP library enriched for large RNAs. The percentage of reads to the total number of reads of the library is plotted for every length size in *black*, and for the reads starting with any the four nucleotides separately: uridine, *blue*; cytosine, *orange*; adenine, *red*; guanine, *light blue*. Note the separation of the piRNA population by a distinct peak at 26–27 nucleotides, but also the overwhelming bias for a uridine at their 5′ends, compared to the longer RNAs. Detailed analysis revealed that a subset of the long RNAs corresponds to intermediate piRNA fragments that have mature 5′ ends with a U bias and untrimmed 3′ ends. Maximum read size of this Illumina run was 76 bases

3. The CLIP libraries prepared from RNAs extracted from the main radioactive signal and higher molecular weights should contain piRNAs, and also fragments of larger RNAs, such as piRNA precursors and target RNAs. These two populations have distinctive characteristics such as genomic origin, nucleotide biases, and size range (*see* **Note 29** and Fig. 2), and therefore downstream bioinformatic analyses should recognize these differences appropriately. Ideally these two populations should be contained in two minimally overlapping peaks in a size distribution plot of the CLIP tags (Fig. 2). The quality of a Piwi CLIP experiment depends on the use of such conditions so that these populations remain faithful to their in vivo profiles. For example, over-digestion of crosslinked RNAs will result in a shift of the average size of piRNAs. Additionally, larger RNAs might be cleaved down to piRNA size, thus "contaminating"

the piRNA population with other RNAs. Since the bioinformatic analysis of the CLIP libraries, and not their preparation and sequencing, is the slowest and most complicated step of a CLIP-Seq project, the best practice is to ensure the highest possible quality of the CLIP library during preparation.

See **Notes 29–31** for further useful information.

4 Notes

1. Preservation of cell viability and integrity is critical for success.
2. Approximately 40 % of tissue wet-weight can be lost during this step. Creating a single cell suspension of the tissue sample before crosslinking leads to further material loss and is not necessary. UV can penetrate small chunks of tissue.
3. In general, 50–60 mg of tissue is sufficient for one CLIP sample in most cases. The sufficient amount of tissue per CLIP sample depends on the expression levels of the RBP in question, and the efficiency of the immunoprecipitation step (Subheading 3.2).
4. Prepare fresh Lysis buffer each time. 1× PXL (1× PBS, 0.1 % SDS, 0.5 % deoxycholate, 0.5 % NP-40) [15] can also be used as the basis of Lysis and Immunoprecipitation buffers, with 5× PXL for subsequent washing steps (5× PBS, 0.1 % SDS, 0.5 % deoxycholate, 0.5 % NP-40). We have observed that most antibodies perform better in 1× PMPG.
5. Cells should be lysed in this step. Hard tissue may require the use of further mechanical means to lyse thoroughly, such as eppendorf pestle.
6. The size of the long crosslinked RNAs should be reduced, so that the protein–RNA complexes can be analyzed by SDS-PAGE. Large complexes will not enter the gel (*see* Subheading 3.4). Furthermore, the purpose of the HITS-CLIP protocol is to identify the binding "site", the exact area of the RNA that is bound by the protein, and therefore areas of long RNAs that are not bound have to be removed. Finally, Illumina deep sequencing is currently limited to 100–120 cycles, and therefore library preparation has to be optimized for this maximum read length. Keep in mind that certain RNases have sequence biases for the cleavage site (RNase T1 cleaves downstream of a G) and any cleaved RNAs will bear this bias.
7. Excess ribonuclease treatment may alter the native size range of the endogenous small RNA populations (piRNAs), and also may lead to the isolation of very small fragments of the long RNAs.

Fragments that are smaller than 19–20 nucleotides cannot be unambiguously mapped on the genome, and this will make bioinformatic analysis of these libraries problematic.

8. Ultracentrifugation "clears" the lysate of insoluble material and higher order complexes of target protein with other possible RBPs, ensuring the immunoprecipitations of single protein–RNA complexes, and therefore the specificity of the RNAs that are isolated at subsequent steps.
9. The addition of "cold" ATP ensures that all RNA adapter molecules have a phosphorylated 5′ end.
10. G-25 Mini spin-column will retain molecules that are smaller than the RL3 adapter (i.e., unincorporated ATP, small adapter fragments) but not the T4 PNK enzyme, which might interfere with subsequent ligation step. Therefore, it has to be heat inactivated. Alternatively, the labeled RNA adapter can be purified by phenol extraction and ethanol precipitation.
11. In order to keep Dynabeads resuspended throughout the course of on-beads enzymatic reactions, a constant mixing motion is required. For the reaction volume and amount of Dynabeads described, this is best achieved on Thermomixer (Eppendorf) set at 1,000 rpm.
12. Addition of excess cold (unlabeled) RL3 ensures that all extracted RNA molecules are ligated with a 3′ adapter and therefore can be amplified by PCR.
13. Nonreducing SDS loading buffer has been reported as an alternative for performing CLIP with proteins with molecular weight close to that of heavy antibody chain, and whose electrophoretic pattern can be affected by the excess of antibody used in the IP. Piwi Argonautes are significantly larger (around 100 kDa) than heavy Ab chains, and no such interference is observed.
14. It is recommended that fresh SDS loading-elution solution is prepared each time. Check for efficient elution by comparing radioactive counts of the eluate and remaining beads. More than 80 % of the counts should be in the eluate.
15. Depending on the size of the protein of interest, you might use different gel concentration and/or running buffer, to achieve optimal resolution at the size range of your protein. Consult the Novex gel migration chart at the Invitrogen website.
16. Wet transfer using Invitrogen's XCell II Blot module has also been used successfully. The iBlot fast transfer apparatus used under manufacturer's instructions for regular western blot is not very efficient for transfer of the crosslinked and labeled RNA–protein complexes, although regular protein is transferred efficiently.

17. Intensifying screens (Fischer) will increase the strength and improve the resolution of the radioactive signal. Keep radiographic cassette in −80 °C during exposure. Bring to room temperature before cutting the membrane.
18. Thin pieces of fluorescent sticker will give sharp signals on the film. This is critical for accurately aligning with the markings on the film after exposure. Proper alignment is a particularly important step, as slight misalignment can lead to incorrect RNA extraction, and confusing results in subsequent PCR amplification steps.
19. For RBPs other than Piwi or Argonautes in general, that may not bind more than one characteristic RNA populations (i.e., small RNAs and mRNAs), the main radioactive signal should contain fragments of all their RNA targets, and membrane fragments from higher positions will only contain larger fragments of the same RNAs.
20. Piwis (and also Agos) form complexes with two different populations of RNAs with different abundances, size and sequence characteristics: piRNAs (small RNAs) and mRNAs, hence the difference in signal intensity and the separation of these complexes by size on the membrane. In the second strategy, cut one membrane fragment containing the main signal but also higher molecular weight smear, as described above. The sample enrichment in the two RNA populations will impact the downstream bioinformatic analysis.
21. The smear extending to higher molecular weights contains longer RNAs. In conditions of high RNase treatment, and depending on the activities of endogenous nucleases during IP steps, the main radioactive signal usually appears sharper but also fainter, and the high molecular weight smear might also be fainter or absent. In conditions of mild RNase treatment (low concentration of RNase T1, or no exogenous nuclease), the signal might appear stronger, more diffuse, and more prominent in higher molecular weights (Fig. 1b). These differences in the radioactive signals correspond to the abundance and the sizes of the RNAs crosslinked with the immunoprecipitated protein, and how these are affected by RNase treatment (high RNase treatment results in smaller RNAs–smaller RNPs and vice versa). Also note that usually there is no band detected on the western blot corresponding to the signals on the autoradiography, and this is because only a small fraction of the total protein in the lysate is crosslinked with RNAs. For mouse Piwi protein CLIP, mild treatment or no exogenous RNase was used. Importantly, the specific radioactive signal is lost in non-crosslinked control samples (Fig. 1c).

22. Rarely, residual organic phase will be visible after centrifugation, and can interfere with the acquisition of the RNA pellet. Remove most of the aqueous supernatant leaving a small amount at the bottom of the tube containing the residual organic solvent and the RNA pellet. Add 400 μL of RNase-free water and vortex rigorously for 1 min. Re-extract by adding 400 μL of Acid Phenol/$CHCl_3$/Isoamyl alcohol 25:24:1, vortexing for 1 min, spinning at maximum speed (~16,000 × *g*) for 5 min at room temperature and collecting upper phase. Remove traces of phenol by extracting with 400 μL of $CHCl_3$/Isoamyl alcohol 24:1. Vortex, spin, and collect aqueous phase as in previous step. Precipitate RNA by adding to the aqueous phase: 40 μL 3 M Sodium Acetate (pH 5.2), 0.5 μL 5 mg/mL Glycogen, and 1 mL ice-cold EtOH, incubating for 30 min at −80 °C and spinning at max speed for 30 min at 4 °C.
23. At this step the RNA pellets are usually extremely small. Do not over-dry the pellets, as they can become very hard to resuspend. Air-dry or place in 37 °C heating block, while constantly monitoring the evaporation of remaining ethanol. Resuspension is best if the pellet is still damp.
24. Metaphor agarose is an intermediate melting temperature agarose with high-resolution capability. Metaphor gels are very fragile. Handle with extreme care, small cuts propagate very fast. Follow manufacturer's instruction for preparation. High temperatures during electrophoresis will render the gels even more susceptible to damage, therefore regulate voltage so that running buffer is not warm.
25. A smeary appearance of both the small RNA and larger RNA populations is preferable at this stage. If you see discrete bands, especially in the sample containing larger RNAs, it may be that only a small number of RNAs were extracted and over-amplified during PCR. Such cDNA libraries will be very poor, if not completely junk.
26. For gel extraction, follow Qiagen's instructions. Weigh each sample's gel fragment. For gel solubilization, add to each tube three times the gel weight (in mg) of QG solubilization buffer (in μL).
27. Quality control and quantification of the piRNA library may be performed by electrophoresis of a small aliquot of the purified library on an Agilent 2100 Bioanalyzer by following the manufacturer's protocol. This step is typically an integral component of the quality control that core facilities perform prior to Illumina sequencing. The protocol may also be modified to incorporate adapters with barcodes for multiplexing.

28. Depending on the length of the CLIP tag and the maximum read length, a variable part of the 3′ adapter can be present within the read, so the trimming approach has to account for this.
29. piRNAs have distinct biases in their sequence (more than 80 % start with a Uridine), their size (25–30 nucleotides long; 26–27 nucleotides for Mili bound piRNAs, and 30 nucleotides for Miwi bound piRNAs) and in their genomic origin (the vast majority cluster in characterized intergenic hotspots, forming pre-pachytene and pachytene clusters [20, 21]). Therefore, the piRNAs comprise a distinct CLIP tag population. CLIP samples enriched in piRNAs (RNA extracted from the main radioactive signal) should be highly similar to published piRNA libraries from IP experiments using standard methodology [4, 10, 18, 21, 22].
30. Recently it was shown that in vivo RNA–protein crosslinking, and subsequent immunoprecipitation, protein degradation, RNA extraction, and reverse transcription of the RNA as described in this technical protocol, introduces characteristic deletions in the deep sequencing CLIP tag reads, which are due to small protein fragments still attached to the extracted crosslinked RNA that the Reverse Transcriptase often skips without incorporating a cognate nucleotide in the cDNA chain. Hence, the *C*rosslink *I*nduced *M*utation *S*ite analysis (CIMS analysis, [16]), reveals RNAs and exact sequences on these RNAs that were bound by the "CLIPed" protein in vivo. The researcher following this protocol should verify the RNA targets of interest by identifying such deletions. We have noticed that a certain sequence bias occurs around CIMS (usually AAA, at −1 to +1 positions around CIMS, also described in ref. [23]), and therefore caution is advised for utilizing CIMS for sequence motif analysis in the protein binding sites.
31. CLIP tags representing fragments of large RNAs (33 nucleotides and up) bound on Piwi proteins should be expected to fall mainly in three categories, based on their genomic origin: piRNA precursors, retrotransposons, and mRNAs [18]. These three categories represent three separate areas of research interest. Note that these categories are not mutually exclusive, i.e., certain mRNAs are processed into piRNAs and also mRNAs contain embedded retrotransposon-derived sequences. Transcripts that are processed into piRNAs should have CLIP tags from both piRNA and large RNA populations (the latter representing unprocessed fragments or intermediate piRNA fragments) aligned into them in sense orientation. Moreover, intermediate fragments of piRNA processing should bear the hallmarks of piRNA processing, i.e., coincidence of their 5′

ends with 5′ ends of mature piRNAs, and therefore increased bias for a Uridine at the 5′ as well [18]. This analysis requires the identification of the exact genomic coordinates of each CLIP tag, and therefore can only be performed for CLIP tags that align only once in the genome (uniquely mapped tags).

Acknowledgment

Supported by NIH grants GM072777 and NS056070 to Z.M.

References

1. Siomi MC, Sato K, Pezic D, Aravin AA (2011) PIWI-interacting small RNAs: the vanguard of genome defence. Nat Rev Mol Cell Biol 12:246–258
2. Juliano C, Wang J, Lin H (2011) Uniting germline and stem cells: the function of Piwi proteins and the piRNA pathway in diverse organisms. Ann Rev Genet 45:447–469
3. Girard A, Sachidanandam R, Hannon GJ, Carmell MA (2006) A germline-specific class of small RNAs binds mammalian Piwi proteins. Nature 442:199–202
4. Aravin A, Gaidatzis D, Pfeffer S, Lagos-Quintana M, Landgraf P, Iovino N, Morris P, Brownstein MJ, Kuramochi-Miyagawa S, Nakano T, Chien M, Russo JJ, Ju J, Sheridan R, Sander C, Zavolan M, Tuschl T (2006) A novel class of small RNAs bind to MILI protein in mouse testes. Nature 442:203–207
5. Lau NC, Seto AG, Kim J, Kuramochi-Miyagawa S, Nakano T, Bartel DP, Kingston RE (2006) Characterization of the piRNA complex from rat testes. Science 313:363–367
6. Kuramochi-Miyagawa S, Kimura T, Ijiri TW, Isobe T, Asada N, Fujita Y, Ikawa M, Iwai N, Okabe M, Deng W, Lin H, Matsuda Y, Nakano T (2004) Mili, a mammalian member of piwi family gene, is essential for spermatogenesis. Development 131:839–849
7. Deng W, Lin H (2002) miwi, a murine homolog of piwi, encodes a cytoplasmic protein essential for spermatogenesis. Dev Cell 2:819–830
8. Vagin VV, Wohlschlegel J, Qu J, Jonsson Z, Huang X, Chuma S, Girard A, Sachidanandam R, Hannon GJ, Aravin AA (2009) Proteomic analysis of murine Piwi proteins reveals a role for arginine methylation in specifying interaction with Tudor family members. Genes Dev 23:1749–1762
9. Kirino Y, Vourekas A, Sayed N, de Lima Alves F, Thomson T, Lasko P, Rappsilber J, Jongens TA, Mourelatos Z (2010) Arginine methylation of Aubergine mediates Tudor binding and germ plasm localization. RNA 16:70–78
10. Reuter M, Chuma S, Tanaka T, Franz T, Stark A, Pillai RS (2009) Loss of the Mili-interacting Tudor domain-containing protein-1 activates transposons and alters the Mili-associated small RNA profile. Nat Struct Mol Biol 16:639–646
11. Mathioudakis N, Palencia A, Kadlec J, Round A, Tripsianes K, Sattler M, Pillai RS, Cusack S (2012) The multiple Tudor domain-containing protein TDRD1 is a molecular scaffold for mouse Piwi proteins and piRNA biogenesis factors. RNA 18:2056–2072
12. Tanaka SS, Toyooka Y, Akasu R, Katoh-Fukui Y, Nakahara Y, Suzuki R, Yokoyama M, Noce T (2000) The mouse homolog of Drosophila Vasa is required for the development of male germ cells. Genes Dev 14:841–853
13. Kuramochi-Miyagawa S, Watanabe T, Gotoh K, Takamatsu K, Chuma S, Kojima-Kita K, Shiromoto Y, Asada N, Toyoda A, Fujiyama A, Totoki Y, Shibata T, Kimura T, Nakatsuji N, Noce T, Sasaki H, Nakano T (2010) MVH in piRNA processing and gene silencing of retrotransposons. Genes Dev 24:887–892
14. Kotaja N, Sassone-Corsi P (2007) The chromatoid body: a germ-cell-specific RNA-processing centre. Nat Rev Mol Cell Biol 8:85–90
15. Chi SW, Zang JB, Mele A, Darnell RB (2009) Argonaute HITS-CLIP decodes microRNA-mRNA interaction maps. Nature 460: 479–486
16. Zhang C, Darnell RB (2011) Mapping in vivo protein-RNA interactions at single-nucleotide resolution from HITS-CLIP data. Nat Biotechnol 29:607–614
17. Licatalosi DD, Mele A, Fak JJ, Ule J, Kayikci M, Chi SW, Clark TA, Schweitzer AC, Blume JE, Wang X, Darnell JC, Darnell RB (2008) HITS-CLIP yields genome-wide insights into brain alternative RNA processing. Nature 456:464–469

18. Vourekas A, Zheng Q, Alexiou P, Maragkakis M, Kirino Y, Gregory BD, Mourelatos Z (2012) Mili and Miwi target RNA repertoire reveals piRNA biogenesis and function of Miwi in spermiogenesis. Nat Struct Mol Biol 19:773–781
19. Kirino Y, Vourekas A, Khandros E, Mourelatos Z, Hobman TC, Duchaine TF (2011) Immunoprecipitation of piRNPs and directional, next generation sequencing of piRNAs. Methods Mol Biol 725:281–293
20. Aravin AA, Sachidanandam R, Bourc'his D, Schaefer C, Pezic D, Toth KF, Bestor T, Hannon GJ (2008) A piRNA pathway primed by individual transposons is linked to de novo DNA methylation in mice. Mol Cell 31:785–799
21. Aravin AA, Sachidanandam R, Girard A, Fejes-Toth K, Hannon GJ (2007) Developmentally regulated piRNA clusters implicate MILI in transposon control. Science 316:744–747
22. Reuter M, Berninger P, Chuma S, Shah H, Hosokawa M, Funaya C, Antony C, Sachidanandam R, Pillai RS (2011) Miwi catalysis is required for piRNA amplification-independent LINE1 transposon silencing. Nature 480:264–267
23. Sugimoto Y, König J, Hussain S, Zupan B, Curk T, Frye M, Ule J (2012) Analysis of CLIP and iCLIP methods for nucleotide-resolution studies of protein-RNA interactions. Genome Biol 13:R67

Chapter 8

DNA Methylation in Mouse Testes

Satomi Kuramochi-Miyagawa, Kanako Kita-Kojima, Yusuke Shiromoto, Daisuke Ito, Hirotaka Koshima, and Toru Nakano

Abstract

DNA methylation of retrotransposons and imprinted genes is accurately regulated in spermatogenesis. In particular, CpG methylation of long interspersed elements-1 (LINE1 or L1) and intracisternal A-particle (IAP) retrotransposons during spermatogenesis has been well characterized. CpG methylation of the regulatory regions of retrotransposons is acquired during embryonic testis development; however, reductions of DNA methylation in LINE1 and/or IAP and/or Rasgrf1, which is an imprinted gene, are observed in deficient mice of piRNA biogenesis concerning. Here, we describe two methods, bisulfite sequencing and Southern blotting using a methylation-sensitive restriction enzyme, for analysis of DNA methylation of LINE1, IAP, and imprinted genes in mouse testes.

Key words Bisulfite sequencing, Southern blotting, Methylation-sensitive restriction enzyme, LINE1, IAP, Imprinted gene

1 Introduction

During spermatogenesis, the DNA methylation status of the regulatory regions in retrotransposons and imprinted genes changes dynamically [1]. These regions are demethylated in primordial germ cells (PGCs) around E10.5–12.5, and reacquisition of DNA methylation (i.e., de novo methylation) takes place in cell cycle arrested prospermatogonia (i.e., gonocytes) in the fetal testis around E15.5–18.5. The high methylation status is retained after birth [2]. piRNAs corresponding to retrotransposons are thought to involve DNA methylation as a guide to sites of de novo methylation in the nucleus. Hence, DNA methylation analysis of two retrotransposon genes, LINE1 and IAP, in germ cells of mutant mouse testes has become the criterion for determining whether a gene involves DNA methylation through piRNAs [3–6]. Additionally, in three paternally imprinted genes, de novo methylation of the differentially methylated region (DMR) of only *Rasgrf1*, but not *H19* or *Dlk1/Gtl2*, is affected by the piRNA pathway [7].

Mikiko C. Siomi (ed.), *PIWI-Interacting RNAs: Methods and Protocols*, Methods in Molecular Biology, vol. 1093, DOI 10.1007/978-1-62703-694-8_8, © Springer Science+Business Media, LLC 2014

Here we describe (1) bisulfite sequencing and (2) methylation-sensitive Southern blotting as two DNA methylation analysis techniques. Bisulfite sequencing involves bisulfite treatment of DNA to determine its methylation pattern. Treatment of DNA with bisulfite converts cytosine residues to uracil, but leaves 5-methylcytosine residues unaffected. Thus, bisulfite treatment introduces specific changes in the DNA sequence that depend on the methylation status of individual cytosine residues, yielding single-nucleotide resolution information about the methylation status of a target DNA strand. In the testis, there are somatic cells as well as germ cells; in addition, LINE1 and IAP are highly methylated in somatic cells. Separating germ cells from somatic cells is of utmost importance; in other words, even a small number of somatic cells disturbs the accurate determination of methylation status with bisulfite sequencing.

In contrast, methylation-sensitive Southern blotting is a classic method of methylation analysis based on the inability of some restriction enzymes to cut methylated DNA. An enzyme pair with CG sequences within their restriction sites, *Hpa*II–*Msp*I (CCGG) is used for this analysis. LINE1 has repetitive sequences in the 5′ noncoding region, which has many CpG sites and acts as a regulatory region (Fig. 1). Usually, these CpG sites are highly methylated in somatic cells and after de novo methylation in germ cells. Therefore, CCGG sites in the regulatory region of LINE1 are not cleaved by the *Hpa*II methylation-sensitive enzyme. Only unmethylated CpG sites which were affected by piRNA-mediated de novo methylation in germ cells are detected as a band of LINE1, hence there is no need to isolate germ cells from somatic cells. Because there are plenty of CCGG sites in the regulatory region of LINE1 in the mouse, it is relatively easy to detect methylation

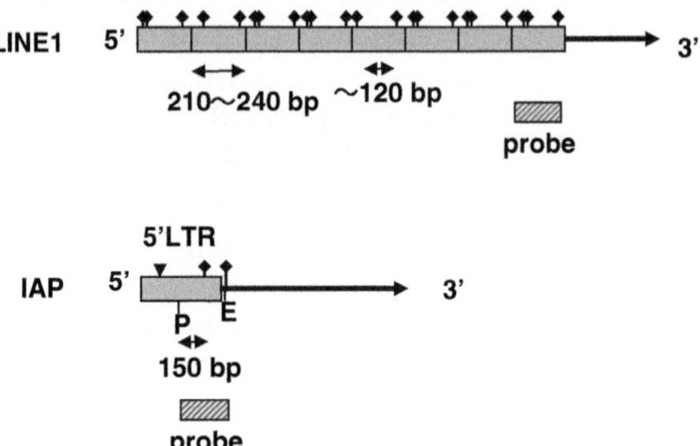

Fig. 1 Schematic structure of the 5′ regulatory region of LINE1 and IAP. LINE1 has a repetitive region of 210–240 bp in the 5′ noncoding region that has many CpG sites. There is one CCGG site in the 5′ LTR of IAP. ⬆ indicates the CCGG site. *Filled inverted triangle* indicates the TGCGCA (*Fsp*I) site

using this Southern blotting method. However, there is one *Hpa*II–*Msp*I (CCGG) site in the 5′ LTR in IAP (Fig. 1), requiring the restriction enzyme to be used in combination with *Pst*I and *Hpa*II/*Msp*I. Alternatively, we recommend the combination of *Eco*RI with *Fsp*I (TGCGCA) as the methylation-sensitive enzyme. In any case, since much fewer CG sites are affected by the piRNA pathway in IAP than in LINE1, the band of unmethylated IAP is harder to detect (*see* **Note 1**).

2 Materials

2.1 Isolation of Germ Cells

For bisulfite sequencing, the isolation of germ cells with high purity is very important. The most certain way to purify cells from mouse testis is to perform a cross with OCT4-GFP transgenic mouse and the identify germ cells as GFP-positive cells by FACS sorting. First, obtain a OCT4-GFP mouse and cross it to the target mice.

If an OCT4-GFP mouse is unavailable, germ cells after birth can be collected with an antibody to EpCAM (Epithelial cell adhesion molecule) [8] or E-cadherin [9], a germ-cell-specific surface antigen by means of FACS sorting or magnetic beads (Fig. 2). When using magnetic beads method, the purity of germ cells can be lower than by FACS sorting and requires considerable attention.

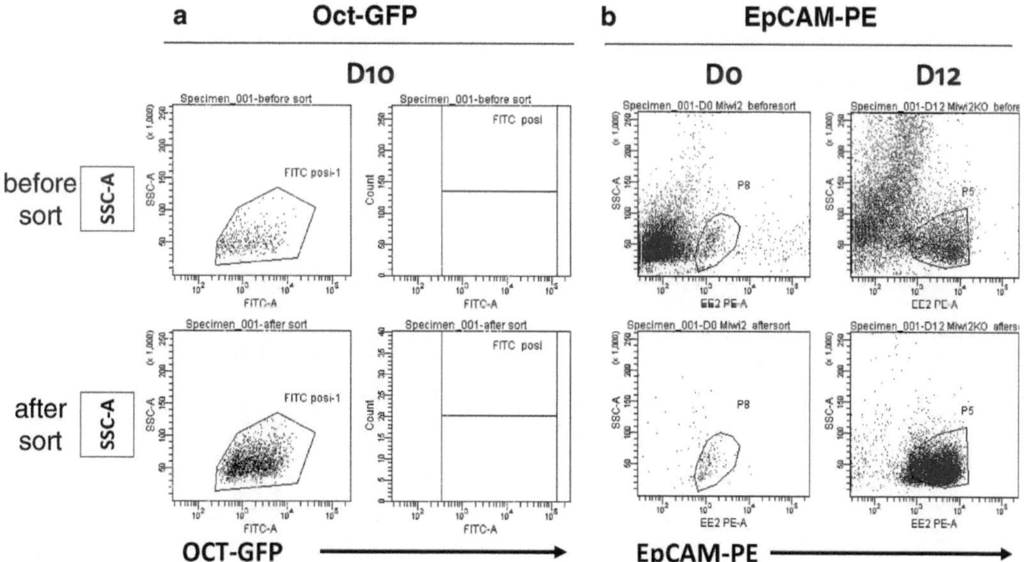

Fig. 2 Flow cytometry analysis before and after sorting. (**a**) The Oct-GFP (FITC) positive germ cells were isolated from the day 10 testis after birth of Oct-GFP mouse by FACSAria (BD). (**b**) The PE-EpCAM positive germ cells were isolated from the day 1 or day 12 testes after birth by FACSAria (BD)

1. Hank's Stock Solutions (HBSS).
2. 0.25 % Trypsin–10 mM EDTA in PBS.
3. 50 mg/ml collagenase in PBS (×50).
4. 10 mg/ml DNase I (Worthington DNaseI for tissue culture and cell isolation; DP) in HBSS (×50).
5. 10 % FCS/medium (any).
6. Nylon mesh 45 μm in pore size or cell strainer (Polystyrene Round Bottle Tube with Cell Strainer Cap; BD Falcon 352235).
7. Rat anti-mouse EpCAM antibody (clone G8.8; BioLegend) or PE anti-mouse CD326 (Ep-CAM) (BioLegend).
8. Dynabeads M-450 conjugated with sheep anti-rat IgG (Dynal Biotech ASA, Oslo, Norway).

2.2 DNA Purification

1. DNA lysis buffer: 1× SSC, 10 mM Tris–HCl (pH 7.5), 1 mM EDTA, 1 % SDS, 1 mg/ml proteinase K, 20 μg/μl RNase A.
2. Phenol/chloroform.
3. Ethanol.

2.3 Bisulfite Sequencing

1. Prepare a bisulfite conversion and DNA cleanup kit; such kits are available from several manufacturers.
2. Prepare the primer sets for bisulfite sequencing listed below.

Primers	Sequence (5′–3′)
H19 (GenBank acc. no. U19619), product size: 423 bp	
First (outside) forward:	GAGTATTTAGGAGGTATAAGAATT
First (outside) reverse:	ATCAAAAACTAACATAAACCCCT
Second (inside) forward:	GTAAGGAGATTATGTTTATTTTTGG
Second (inside) reverse:	CCTCATTAATCCCATAACTAT
Rasgrf1 (GenBank acc. no. AF021791), product size: 249 bp	
First forward:	GAGAGTATGTAAAGTTAGAGTTGTGTTG
Second forward:	TAAAGATAGTTTAGATATGGAATTTTGGG
First and Second reverse:	ATAATACAACAACAACAATAACAATC
Line1(Tf) (GenBank acc. no. D84391), product size: ~270 bp	
Forward:	GTTAGAGAATTTGATAGTTTTTGGAATAGG
Reverse:	CCAAAACAAAACCTTTCTCAAACACTATAT
Line1(A) (GenBank acc. no. M13002), product size: ~310 bp	
Forward:	TTATTTTGATAGTAGAGTT
Reverse:	CAAACCAAACTCCTAACAA
IAP, product size: 259 bp (ch3)	
Ch3-first forward:	GTTTGTAATGGTGGGAGAT
First reverse:	AAATAAATTGTGGGAAGT
Second forward:	TTGTGTTTTAAGTTGGTAAATAAATAATT
Second reverse:	CAAAAAAAACACCACAAACCAAAAT

2.4 Southern Blotting

1. Restriction enzymes *Hpa*II, *Msp*I, *Pst*I.
2. Agarose gel (1 %)/TAE buffer and TAE running buffer.
3. 50 mM HCl.
4. Denaturing buffer: 0.5 N NaOH, 1.4 M NaCl.
5. Neutralization buffer: 0.5 M Tris–HCl (pH 7.5), 3 M NaCl.
6. Nylon filter membrane (Hybond N+; GE Healthcare).
7. 3 MM filter paper and paper towel or blotting sponge (Blotting Pad; NIPPON Genetics).
8. 20× SSC (3 M NaCl, 0.3 M sodium citrate).
9. Prehybridization and hybridization buffer: 0.2 M $NaHPO_4$ (pH 7.2), 1 mM EDTA, 1 % BSA, 7 % SDS.
10. Preparing LTR region of IAP or the 5′ noncoding region Tf or A type LINE1-specific probes:

    Subclone the PCR products using the primers below.

Line1 (Tf) (GenBank acc. no. D84391, product size: 283 bp),	
Line-1 Tf-F1	TAGGAAATTAGTCTGAACAGGTGAG
Line-1 Tf-R1	TCAGACACTGTGTTGCTTTGGCAG
Line1 (A) (GenBank acc. no. M13002, product size: 1,112 bp)	
Line-1 A-F1	GAGTTTTTGAGTCTGTATCC
Line-1 A-R1	CTCTCCTTAGTTTCAGTGG
IAP (GenBank acc. no. M17551, product size: 222 bp)	
IAP-Pst-F1	CTGCAGCCAATCAGGGAGTG
IAP-Eco-R1	TTTTCTCGTCCCGGAATTCGG

    Sequence the subcloned PCR products and compare their sequences with nucleotides 874–1,156 in D84391, 531–1,642 in M13002, or 148–369 in M17551, respectively. Because the retrotransposon has several sequence variations, the sequences need not be completely identical to D84391, M13002, or M17551. Purify the insert DNA from the vector plasmid. (If you use pGEM-T easy (Promega) as a cloning vector, you can cut out the insert with *Eco*RI.)

11. 2× SSC-0.1 % SDS.
12. 0.2× SSC-0.1 % SDS.

3 Methods

3.1 Isolation of Germ Cells by FACS Sorting Using OCT4-GFP Tg Mice

1. Remove testes from mouse.
2. Immerse the testes in 0.5 ml HBSS containing 10 μl of collagenase (×50) and 10 μl of DNaseI (×50).
3. Incubate at 37 °C for 15 min.

4. Wash with HBSS 1 ml twice (cfg 1,000 × *g* for 2 min) (If using embryonic testes, **steps 2–4** can be omitted).
5. Mince the testes by pipetting in 0.5 ml of 0.25 % Trypsin–10 mM EDTA in PBS.
6. Incubate at 37 °C for 10 min.
7. Add 0.5 ml of 10 % FCS/medium and pipette.
8. Add 20 μl of DNaseI (×50).
9. Incubate at 37 °C for 5 min.
10. Wash with HBSS 1 ml twice (cfg 1,000 × *g* for 2 min).
11. Resuspend the cells in 10 % FCS/medium.
12. Pass the cells through a cell strainer.
13. Sort the GFP-positive cells according to the manual (Fig. 2a).
14. Collect cells by centrifugation.

3.2 Isolation of Germ Cells by FACS Sorting Using PE-Conjugated Anti-EpCAM Antibody

1–10. Same above.

11. Resuspend in 0.5 ml of 5 % BSA/HBSS and pass the suspension through a cell strainer (BD Falcon).
12. Add 0.5-μg PE-conjugated anti-EpCAM antibody and incubate for 1 h. at 4 °C with gentle mixing.
13. Wash the cell suspension with HBSS (0.5 ml) three times (cfg 1,000 × *g* for 2 min at 4 °C).
14. Resuspend the cells in 5 % BSA/HBSS.
15. Pass the cells through a cell strainer.
16. Sort the PE positive cells according to the manual (Fig. 2b).
17. Collect cells by centrifugation.

3.3 Isolation of Germ Cells Using EpCAM-Dynabeads

1–11. Same above.

12. Prepare the antibody-coated Dynabeads.
    Wash Dynabeads M-450 in 0.5 ml of HBSS three times.
    Add 0.5-μg rat anti-mouse antibody against EpCAM to 15 μl of 50 % slurry Dynabeads.
    Rotate at 4 °C for 30 min or overnight before mixing the Dynabeads with cells.
    Wash with HBSS three times.
13. Add a complex of anti-EpCAM antibody and Dynabeads to the cell suspension, and incubate for 45 min at 4 °C with gentle mixing.
14. Separate the EpCAM positive cells with Dynal MPC (Dynal AS, Oslo, Norway) and wash four times for 5 min at 4 °C.

3.4 Isolation of Germ Cells Using E-cadherin-Dynabeads

1. Remove testes from mouse (at least day 14 after birth).
2. Fix the testes with 0.5 ml of 70 % ethanol for 20 s.
3. Wash with PBS 0.5 ml three times.

4. Mince the testes with a surgical knife.
5. Suspend well and treat with 0.5 ml of 1 mg/ml collagenase type 2 in HBSS for 10–30 min at 37 °C.
6. Wash the cell suspension with HBSS (0.5 ml) three times (cfg 2,000 × *g* for 2 min at 4 °C).
7. Resuspend in 0.5 ml of HBSS and pass the suspension through a cell strainer (BD Falcon).
8. Prepare the antibody-coated Dynabeads.
   Wash Dynabeads M-450 in 0.5 ml of HBSS three times.
   Add 0.5-μg rat anti-mouse antibody against E-cadherin to 15 μl of 50 % slurry Dynabeads.
   Rotate at 4 °C for 30 min or overnight before mixing the Dynabeads with cells.
   Wash with HBSS three times.
9. Add a complex of anti-E-cadherin antibody and Dynabeads to the cell suspension, and incubate for 45 min at 4 °C with gentle mixing.
10. Separate the E-cadherin-positive cells with Dynal MPC (Dynal AS, Oslo, Norway) and wash four times for 5 min at 4 °C.

3.5 DNA Purification and Bisulfite Sequencing

1. Add 0.5 ml of DNA lysis buffer and incubate overnight at 56 °C.
2. Extract sample DNA according to general protocol.
3. Treat the sample DNA with a bisulfite conversion kit following the protocol.
4. Bisulfite analysis: First, check the purity of germ cells by analyzing the *H19* imprinted gene. Because *H19* is paternally imprinted genes, if germ cells were purified completely, then all CpGs in the DMR of H19 will be methylated. Somatic cells have paternal and maternal alleles; in other words, the methylation ratio of CpG in DMR of imprinted genes is 50 %. This means that 90 % methylation of H19 indicates 20 % contamination of somatic cells (ratio of germ cells is 80 %) (Fig. 3).

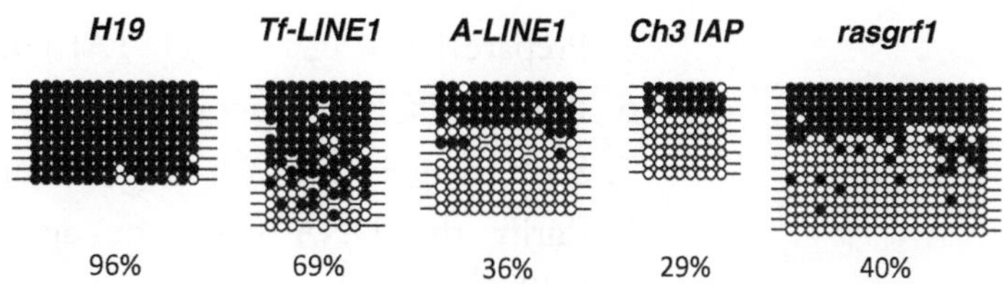

Fig. 3 Analysis of bisulfite sequencing. *Filled circle* and *opened circle* represent methylated and unmethylated CpG, respectively. CpG methylation status of H19, Tf type Line1, A type Line1, 1Δ1 IAP in Ch3 and Rasgrf1 in testicular germ cells from *Mili* deficient mice on day 12 after birth (*see* **Note 2**)

If the purity of germ cells is not good (e.g., methylation levels below 95 % for H19), the sample should not be used for further bisulfite analysis.

PCR reaction	
Reaction mixture	
Bisulfite-treated DNA	1–2 μl
10× PCR buffer (+Mg^{2+})	3 μl
2.5 mM dNTPs	2.4 μl
Forward 10 μM primer	3 μl
Reverse 10 μM primer	3 μl
[a]Taq polymerase	0.3–0.5 μl
H_2O to 30 μl	

[a]Taq polymerase: Use a Taq DNA polymerase that preferentially adds an adenine to the 3′ end of the product for TA cloning. Subcloning by TA cloning is easier and quicker

H19, Rasgrf1	
First	94 °C for 30 s
	56 °C for 60 s
	72 °C for 30 s × 30 cycles
Second	94 °C for 30 s
	56 °C for 30 s
	72 °C for 30 s × 30 cycles
LINE1	
	94 °C for 30 s
	50 °C for 30 s
	72 °C for 30 s × 30 cycles
IAP	
First	94 °C for 30 s
	50 °C for 60 s
	72 °C for 30 s × 30 cycles
Second	94 °C for 30 s
	51 °C for 30 s
	72 °C for 30 s × 30 cycles

5. Prepare 1.5 % agarose gel (TAE) in a gel apparatus.
6. Run the PCR product in 1× TAE buffer.
7. Stain the gel with EtBr and cut out the DNA band.
8. Purify the DNA from the agarose gel according to the manual.
9. Mix PCR product and TA-cloning vector (pGEMTeasy; Promega), 10× ligation buffer and T4 ligase (25–50 ng TA-cloning vector is enough).

10. Incubate at 16 °C for 30–120 min.
11. Transfer the ligated vector to LB plates. For blue/white color selection, LB plates should be spread with 5-bromo-4-chloro-3-indolyl-β-D-galactoside (X-Gal), and sopropylthio-β-D-GALACTOSIDE (IPTG) before transformation.
12. Randomly pick white colonies and prepare the plasmid DNA.
13. Sequence (regarding the primer for bisulfite sequencing, SP6 or T3 often gives better results than T7) and compare the sequences to the reference sequence (*see* **Note 3**).

3.6 Methylation-Sensitive Southern Blotting

1. Collect the testis.
2. Add 0.5 ml of DNA lysis buffer and incubate at 56 °C overnight.
3. Extract sample DNA according to general protocol.
4. Mix 5 μg (3 μg for mini gel) sample DNA, 2 μl of 10× reaction buffer, and H_2O to 18 μl.
5. Add 2 μl of *Pst*I (10 U/μl) and incubate for more than 2 h at 37 °C.
6. Add 50 μl of ethanol and centrifuge at maximum speed.
7. Wash with 70 % ethanol and dry.
8. Resuspend DNA in 12.4 μl of H_2O, and add 1.8 μl of 10× reaction buffer, 1.8 μl of 10× BSA, and 2 μl of *Hpa*II or *Msp*I (10 U/μl).

   (For detecting Line1, **steps 4–7** can be omitted. You may digest 5 μg (or 3 μg) DNA with *Hpa*II or *Mst*I directly).
9. Incubate overnight at 37 °C
10. Prepare 1.0 % agarose gel (TAE) in the gel apparatus.
11. Mix digestion with 2 μl of 10× loading buffer.
12. Run gel electrophoresis in 1× TAE buffer.
13. Stain the gel with EtBr and confirm sufficient digestion on UV transilluminator.
14. Soak the gel in 200 ml (100 ml for mini gel) of 50 mM HCl in a tray for 10 min at room temperature.
15. Decant acid and then rinse gel with water.
16. Soak the gel in 200 ml of denaturation buffer for 15 min at room temperature.
17. Decant the buffer and soak with another 200 ml of denaturation buffer for 30 min at room temperature.
18. Soak the gel in 200 ml of neutralization buffer for 15 min and another 30 min at room temperature.
19. Transfer DNA to a nylon membrane overnight.
20. Fix the DNA to the filter by UV-irradiation.

3.7 Hybridization

1. Prehybridize the filter in buffer at 65 °C for more than 2 h in a sealed container or glass hybridization bottle.
2. Label the probe with ^{32}P-CTP according to the manual (labeling kits are available from several manufacturers).
3. Discard the prehybridization buffer and add new hybridization buffer.
4. Add ^{32}P-labeled DNA probe to the sealed container or glass hybridization bottle.
5. Incubate at 65 °C overnight.
6. Discard the hybridization buffer.
7. Wash the filter with 2× SSC-0.1 % at 65 °C for 15 min.
8. Wash the filter with 2× SSC-0.1 % at 65 °C for 30 min twice.
9. Wrap the filter with plastic wrap.
10. Expose the filter with X-ray film or an IP plate.
11. Develop the film.

4 Notes

1. IAP has many deletion mutants and only one subtype (1Δ1 type) is affected by the piRNA pathway. Because the 1Δ1 type IAP represents only ~5 % of all IAP and all IAP deletion types have a 5′ LTR, it is not easier to detect the band of unmethylated IAP compared to the that of unmethylated LINE1 in southern blotting. Moreover, when using bisulfite sequencing, it is necessary to examine the CpG sites in the 5′ LTR of individual 1Δ1 IAP sites not the total 5′ LTR of IAP (Figs. 4 and 5).
2. Because there are plenty of LINE1 (Tf and A) sites in the genome and each LINE1 sequence has some mutations even at the CpG sites, the bisulfited sequencing samples often do not have the complete match with reference sequence (*see* Fig. 3; Tf-LINE1 and A-LINE1).
3. Reference sequences for bisulfite sequence (underline; First primer, double underline; Second primer, shaded text; CpG sequence)

<H19>

GAGCATCCAGGAGGCATAAGAATTCTGCAAGGAGAC
CATGCCCTATTCTTGGACGTCTGCTGAATCAGTT
GTGGGGTTTATACGCGGGAGTTGCCGCGTGGT
GGCAGCAAAATCGATTGCGCCAAACCTAAAGA
GCCCCCCCACCCCTGGTATTGGAATTCACAAATG
GCAATGCTGTGGGTCACCCAAGTTCAGTACCTCAGGG
GGGTCACAAATGCCACTAGGGGGGCAGGACACAT
GCATTTTCTAGGCTGGTACCTCGTGGACTCGGACTC
CCAAATCAACAAGGTCGGCTTACTCTCTGCAAAGAAT
CCTTTGTGTGTAAAGACCAGGGTTGCCGCACGGCG

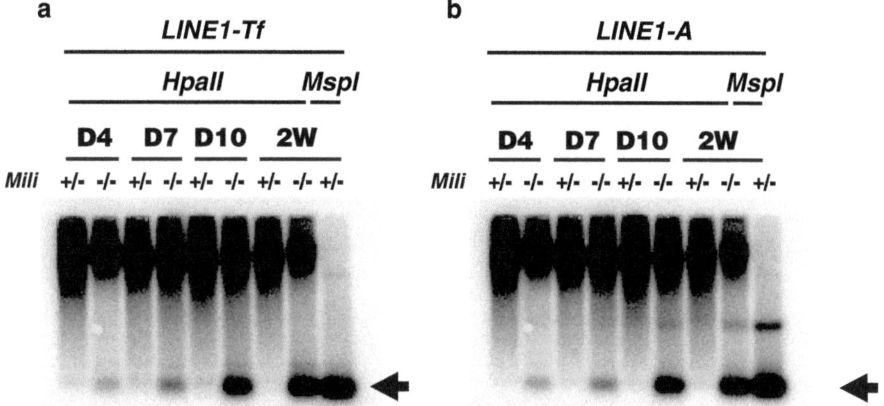

Fig. 4 Methylation-sensitive Southern blot analysis of LINE1 probed with Tf type (**a**) and A type (**b**). Genomic DNA (3 μg) from the whole testis of *Mili* heterozygous (+/–) and deficient (–/–) mice on days 4, 7, 10, and 14 after birth was digested with *Hpa*II or *Msp*I. *Arrows* indicate unmethylated CCGG fragments of the LINE1 regulatory region in *Mili* –/– testes

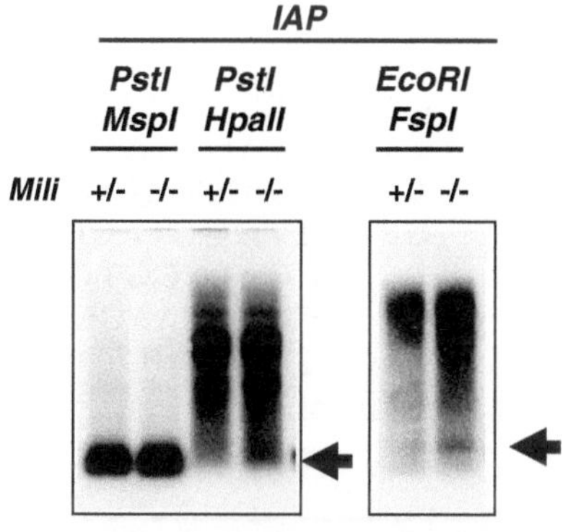

Fig. 5 Methylation-sensitive Southern blot analysis of IAP. Genomic DNA (3 μg) from the whole testis of *Mili* heterozygous (+/–) and deficient (–/–) mice on day 21 after birth was digested with *Pst*I plus *Hpa*II or *Msp*I, or *Eco*RI plus *Fsp*I. *Arrows* indicate unmethylated CG–*Pst*I or *Eco*RI fragments of the IAP LTR region in *Mili* –/– testes

GCAGTGAAGTCTCGTACATCGCAGTCCTAAAACGGAT
TGCAACTGATTGAGTTTTCTCCCCTATCACCATCTATGA
TCCCATAGTCATGGGCTTCATGAGGCCAGGGGTTC
ATGCTAGTCCTTGAT
<LINE1 (Tf)>
GCCAGAGAACCTGACAGCTTCTGGAACAGGCAGAAGCA
CAGAGGGGCTGAGGCAGCACCCTGTGTGGGCCGGGGA
CAGCCGGCCACCTTCCGGACCGGAGGACAGGTGCCCACCCG
GCTGGGGAGGCGGCCTAAGCCACAGCAGCAGCGG
TCGCCATCTTGGTCCCGGGACTCCAAGGAACTTAGGAAT
TTAGTCTGCTTAAGTGAGAGTCTGTACCACCTGGGAACT
GCCAAAGCAACACAGTGTCTGAGAAAGGTCCTGTT
TTGG

<LINE1 (A)>
CTACCTTGACAGCAGAGTCTTGCCCAACACCCGCAAGGG
CCCACACGGGACTCCCCACGGGACCCTAAGACCTCTGG
TGAGTGGAACACAGCGCCTACCCCAATCCAATCGCGTGGAACT
TGAGACTGCGGTACATAGGGAAGCAGGCTACCCGGGC
TTGATCTGGGGCACAAACCCCTTCCACTCCACTCGAGCCC
CGGCTACCTTGCCAGCTGAGTCGCCTGACACCCGCAAGGGC
CCACACAGGATTCCACACGTGATCCTAAGACCTCTAGTGAG
TGGAACACAACTTCTGCCAGGAGTCTGGTTCG

<IAP (Ch3)>
GTCTGTAATGGTGGGAGACTAATTAAAAACAAACTGTGGG
AAGCCGCCCCCACATTCGCCGTCACAAGATGGCGCTGA
CATCCTGTGTTCTAAGTTGGTAAACAAATAATCTG
CGCATGAGCCAAGGGTATTTACGACCACTTGTACTCTGTTT
TTCCCGTGAACGTCAGCTCGGCCATGGGCTGCAGCCAAT
CAGGGAGTGATGCGCCCTAGGCAATGGTTGTTCTCTTTAAA
GAGGGAAGGGGTTTTCGTTTTCTCTCTCTTGCTTCTTGCT
CTCTTTTCCTGAAGATGTAAGAATAAAGCTTTGTCGCAGAA
GATTCTGGTCTGTGGTGTTCTTCCTGGCCGGTCGTGAGAA
CGCGTCGAATAACAATTGGTGCCGAATTCCGGGACGAG
AAAATCCGGGACGAGAAAAAAACTCCGGACTGGCGCAGG
AGGGATACCTCATTCCAGAAC

<Rasgrf1>
GAGAGTATGTAAAGCCAGAGCTGTGCTGCTGCCGCTAAAGA
TAGTTTAGATATGGAATTCTGGGGACTCTTCAGAGAGT
TTATAAAGCCAGCGCTGTGCTGCTGCCGCGCTTCGCGGCT
GCACTTCGCTACCGTTTCGCGGCCGCGCTGCTGCTCCC
ACATCCATCCGTGGCTACCGCTATTGCTGTTGCTGCACC
GCTGCCGCTAAGCTATGGCTGCCGCACTTCACTGTTGCGC
TACCGCTGCGCTACAACTACCACGACTGCTACTGCTGCTG
CTGCACTAC

Acknowledgments

The author thanks Ms. N. Asada for technical support. This work was supported in part by grants from the Japan Society for the Promotion of Science, the Ministry of Education, Science, Sports, and Culture.

References

1. Lane N, Dean W, Erhardt S, Hajkova P, Surani A, Walter J, Reik W (2003) Resistance of IAPs to methylation reprogramming may provide a mechanism for epigenetic inheritance in the mouse. Genesis 35:88–93
2. Sasaki H, Matsui Y (2008) Epigenetic events in mammalian germ-cell development: reprogramming and beyond. Nat Rev Genet 9: 129–140
3. Aravin AA, Sachidanandam R, Girard A, Fejes-Toth K, Hannon GJ (2007) Developmentally regulated piRNA clusters implicate MILI in transposon control. Science 316:744–747
4. Carmell MA, Girard A, van de Kant HJ, Bourc'his D, Bestor TH, de Rooij DG, Hannon GJ (2007) MIWI2 is essential for spermatogenesis and repression of transposons in the mouse male germline. Dev Cell 12:503–514

5. Kuramochi-Miyagawa S et al (2008) DNA methylation of retrotransposon genes is regulated by Piwi family members MILI and MIWI2 in murine fetal testes. Genes Dev 22: 908–917
6. Watanabe T et al (2011) MITOPLD is a mitochondrial protein essential for nuage formation and piRNA biogenesis in the mouse germline. Dev Cell 20:364–375
7. Watanabe T et al (2011) Role for piRNAs and noncoding RNA in de novo DNA methylation of the imprinted mouse Rasgrf1 locus. Science 332:848–852
8. Kanatsu-Shinohara M, Takashima S, Ishii K, Shinohara T (2011) Dynamic changes in EPCAM expression during spermatogonial stem cell differentiation in the mouse testis. PLoS One 6:e23663
9. Tokuda M, Kadokawa Y, Kurahashi H, Marunouchi T (2007) CDH1 is a specific marker for undifferentiated spermatogonia in mouse testes. Biol Reprod 76:130–141

Chapter 9

Analysis of Small RNA-Guided Endonuclease Activity in Endogenous Piwi Protein Complexes from Mouse Testes

Michael Reuter and Ramesh S. Pillai

Abstract

Small RNAs associate with members of the Argonaute family to function in gene regulation, transposon control, and creation of silent chromatin domains. In this partnership, small RNAs act as guides for the bound Argonaute and other associated proteins. Complementary base pairing of small RNAs to target nucleic acid molecules allow specificity for the small RNA-mediated functions. One key activity of some Argonaute protein family members is their small RNA-guided endonuclease activity called Slicer action. Here we describe a protocol that can be used to probe slicer activity in endogenous Piwi complexes isolated from mouse testes.

Key words Piwi, Slicer, piRNA, DDH, Miwi

1 Introduction

Small RNAs are defined as RNAs in the size range of 19–30 nucleotides (nt) that are bound by a member of the Argonaute family of RNA-binding proteins [1–3]. They are implicated in silencing roles in controlling cellular gene expression or activity of transposable elements. The basic unit of these silencing pathways is an Argonaute protein bound to a single-stranded small RNA [4]. The small RNA serves as a guide for the bound Argonaute protein, channeling them to nucleic acid targets. Guidance is achieved by complementary base pairing of the small RNA to the target molecule. Three distinct classes of small RNAs are currently identified: microRNAs (miRNAs), small-interfering RNAs (siRNAs), and Piwi-interacting RNAs (piRNAs) [2]. The miRNAs and siRNAs bind the AGO clade of Argonautes which are ubiquitously expressed. In contrast, piRNAs associate with the PIWI clade which is exclusively found in animals and mainly display gonad-specific expression [5]. Multiple Argonautes are represented in genomes of most organisms; the mouse genome has four AGO proteins (Ago1-4) and three PIWI proteins (Mili, Miwi, and Miwi2).

Mikiko C. Siomi (ed.), *PIWI-Interacting RNAs: Methods and Protocols*, Methods in Molecular Biology, vol. 1093,
DOI 10.1007/978-1-62703-694-8_9,

Argonaute proteins can be structurally divided into four domains: N terminus, PAZ (named after proteins Piwi, Argonaute, and Zwille), MID (middle), and PIWI (named after fly protein Piwi) [6]. Generally, small RNAs carry a 5′ monophosphate and a 3′ hydroxyl end. However, exceptions exist. In worms, some small RNAs are generated by direct transcription from RNA templates by RNA-dependent RNA polymerases, resulting in a 5′ triphosphate [7]. On the other hand, all plant small RNAs, endosiRNAs in *Drosophila* and Piwi-interacting RNAs carry a 2′-*O*-methyl modification at the 3′ end [8–11]. The small RNA is held within a groove in the Argonaute protein via 5′ and 3′ end recognition by the MID and PAZ domains, respectively [12–15]. Perhaps the MID domain is the only site where the Argonaute protein displays base-specific recognition of the small RNA. This is evident from the observed bias for a particular nucleotide at the 5′ end of small RNA populations: U in miRNAs and piRNAs, G in 22G-siRNAs from worms, and C or G in distinct plant small RNA populations [16]. Otherwise, most contacts with the small RNA is along the sugar-phosphate backbone resulting in an overall sequence-independent binding [17, 18]. This allows for an enlarged repertoire of guidance information to be invested in Argonaute protein complexes. The RNA is accommodated such that the bases are exposed to the solvent and available for base-pairing interactions with the target.

Argonaute proteins can act as platforms for assembly or recruitment of factors that result in silencing of the nucleic acid target. The various mechanisms include translational repression, mRNA decay by deadenylation and transcriptional silencing by histone modifications or DNA methylation [19–23]. Perhaps a much conserved activity of Argonaute complexes is the small RNA-guided endonuclease activity called "Slicing" or "Slicer activity" (as target nucleic acid is sliced into two fragments) [24–28]. In fact, in organisms where multiple Argonautes are present, at least one is a Slicer, while those with a single Argonaute always display the slicer activity. Crystal structures of Argonaute proteins revealed that the PIWI domain adopts a fold similar to RNaseH, an endoribonuclease that cleaves the RNA in an RNA: DNA hybrid [27]. Indeed, a bacterial Argonaute (*A. aeolicus* Ago) uses DNA as a guide to cleave RNA targets [29]. However, in most other organisms a small RNA is used to guide cleavage of RNA targets. Slicer cleavage is a very precise magnesium-dependent biochemical reaction that results in cleavage of target RNA at the scissile phosphate group opposite to the phosphate between the 10th and 11th nucleotide of the small RNA (counting from the 5′ end of the small RNA) [30]. The product of slicer action is the generation of two target RNA fragments that carry a 5′ phosphate and a 3′ hydroxyl group at the site of cleavage.

One consequence of Argonaute slicer activity is the destruction of the target RNA. Argonaute-mediated endonucleolytic cleavage of an RNA target opens up access to 5′–3′ exonuclease activity of Xrn1 and 3′–5′ exonuclease action of the exosome, resulting in its complete destruction [31]. In animal gonads, piRNA-guided Piwi slicer action is required to destroy target transposon transcripts [32–36]. However, slicer action of Argonautes does not always lead to destruction of the target. For example, Piwi slicer activity is central to a piRNA amplification pathway called the Ping-pong cycle or secondary piRNA biogenesis [33, 34]. In these situations, instead of destroying the resulting cleavage fragments, an ill-defined biogenesis machinery allows loading of one of the slicer cleavage fragments (the one carrying the 5′ phosphate) to a new Piwi protein to allow its maturation as a new piRNA. Slicer action of Argonautes is shown to be essential for noncanonical miRNA biogenesis [37, 38] and for RISC formation by cleaving the passenger strand of siRNA duplexes [39]. Finally, slicer activity of fission yeast Ago1 is required for spreading of the small RNA-driven heterochromatic silencing [40].

Several strategies can be employed for detecting slicer activity within cells. First, reporters that have at least a single perfectly complementary binding site for an endogenous small RNA are used to score for the silencing effect. Second, a direct read out slicing events in vivo is cloning and sequencing of slicer cleavage products from total RNA by 5′-rapid amplification of cDNA ends (5′-RACE). Such experiments take advantage of the 5′ phosphate present on one of the slicer products to enable adapter ligation and accurate detection of the cleavage site. The signature 10 nt overlap of the 5′ ends of RACE products and small RNAs is then examined [41]. Third, a physiologically informative approach is the study of organisms carrying slicer-inactivating mutations. Crystal structures of Argonautes identified two Aspartates and one Histidine (DDH motif) as residues that coordinate the Mg^{2+} ion essential for catalysis [27]. Indeed, all demonstrated slicers carry this motif and mutation of any one of these residues (e.g., DDH to ADH) results in catalytically dead Argonautes. Mouse *Ago2*^{ADH} mutants die immediately after birth [38], while *Mili*^{DAH} [36] and *Miwi*^{ADH} [35] mutants display male infertility and activate transposons. However, similar mutations predicted to inactivate slicer activity of *Drosophila* Piwi [42], *C. elegans* Prg-1 (worm Piwi) [43], and mouse Miwi2 [36] had no discernible phenotypes. Thus, a direct approach to examine slicer activity in vitro is desirable. The slicer assay provides a direct proof for the catalytic activity validating or supporting observed mutant phenotypes.

In this chapter, we describe in detail the protocol that can be used for examining slicer activity of mouse Piwi protein, Piwil1 (Miwi), using immunopurified protein complexes.

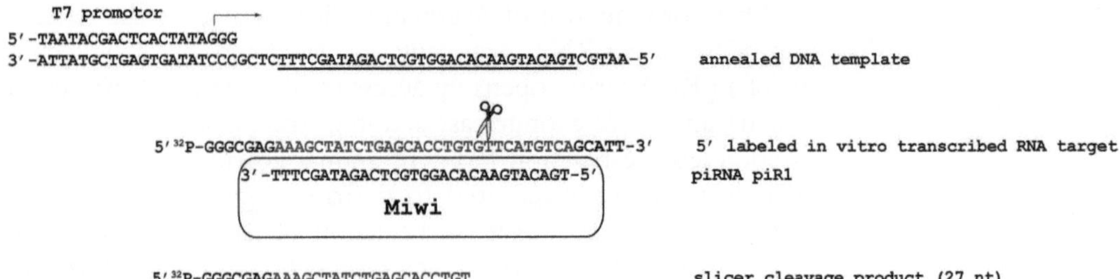

Fig. 1 DNA templates for the in vitro transcription of slicer targets were generated by the annealing of two DNA oligonucleotides creating a double-stranded T7 promoter, with an extended single-stranded region containing the small RNA guide sequence (*underlined*). Slicer cleavage of the target RNA substrate (42 nt) leads to the production of a 27 nt cleavage product

2 Materials

2.1 In Vitro Transcription

1. DNA templates for in vitro transcription were generated from two annealed oligonucleotides as shown in Fig. 1. This created a double-stranded T7 promoter sequence (TAATACGACTCACTATAGGG) with a single-stranded stretch containing the template sequence for the RNA to be transcribed. This part contains of a sequence that is identical to a guide piRNA (piR-A) abundantly present in Miwi complexes (Fig. 1). When transcribed, the RNA substrate will be recognized by the piRNA molecule. DNA oligos were purchased from a commercial supplier (Invitrogen).
2. Site-specific mismatches between guide and substrate RNA were introduced by using template DNA oligonucleotides carrying residues identical to the piRNA sequence at the required positions.
3. Buffers: 10× Annealing buffer: 100 mM Tris/HCl pH 7.5, 500 mM NaCl; 5× transcription buffer: 1 M HEPES pH 7.5, 0.2 M DTT; 5× $MgCl_2$-spermidin buffer: 150 mM $MgCl_2$, 10 mM spermidin.

2.2 Urea-PAGE

1. Buffers: 2× Formamid loading buffer: 80 % Formamid, 1 mM EDTA, 40 ng/ml Bromophenol blue, 40 ng/ml Xylenxyanol; 10× TBE buffer: per 1 l 108 g Tris base, 55 g boric acid, 7.5 g EDTA; Urea-PAGE buffer I: 8.3 M Urea, 20 % acrylamide, *NN*-methylen bis-acrylamide (19:1), 1× TBE; Urea-PAGE buffer II: 8.3 Urea, 1× TBE (Tris, Boric Acid, EDTA).
2. Gel chambers consisted of two glass plates (20 × 20 cm) separated by 1 mm spacers at each side. Chambers were sealed with one layer of package tape at all sides. The loading volume of combs ranged from 10 to 30 μl.

3. Centrifuge tube filters 0.2 μm (spin-X; Costar) were used at 2,000 × *g* for the removal of gel pieces after overnight elution.

2.3 Immunoprecipitations

1. Buffers: Wash buffer: 25 mM Tris/HCl pH 8.0, 150 mM NaCl, 1 mM $MgCl_2$, 0.01 % NP40; Testes extract buffer: 10 % glycerol, 50 mM Tris/HCl pH 8.0, 150 mM NaCl, 5 mM $MgCl_2$, 0.5 % Na deoxycholate, 1 % Triton-X 100, 1× proteinase inhibitors (Complete EDTA-free protease inhibitors; Roche), 10 μg/ml yeast tRNA (Fermentas) and 1× Vanadyl ribonucleoside complex (as RNase inhibitor; Sigma).
2. All immune precipitations were performed with protein G Sepharose (Amersham) in a batch purification mode (volumes of 1–10 ml) on ice using the appropriate centrifuges at 4 °C and 600 × *g*.

2.4 Northern Blotting

1. Proteinase K buffer (10×): 100 mM Tris/HCl pH 7.5, 5 % SDS (sodium dodecyl sulfate), 50 mM EDTA
2. Church buffer: 0.25 M Na phosphate pH 7.2 (prepare 1 M Na_2HPO_4 (A), 1 M NaH_2PO_4 (B). To obtain 1 L of 0.5 M Na phosphate pH 7.2, mix 342 ml of A + 158 ml of B + 500 ml H_2O), 1 mM EDTA, 1% BSA (bovine serum albumin), 7 % SDS.
3. Membranes were from Amersham (HybondN+). Before the transfer, membranes were washed first in water and then in 0.5× TBE buffer.
4. Typhoon Scanner (GE Health) or other phosphor imaging apparatus. Image Quant Analysis software. An alternative possibility is to expose radioactive blots to X-ray films (GE Health).

2.5 Slicer Assay

1. Slicer assay buffer: 10 mM Tris/HCl pH 7.5, 100 mM KCl, 2 mM $MgCl_2$, 40 U RiboLock RNase inhibitor (Fermentas).

3 Methods

3.1 In Vitro Transcription to Generate RNA Substrate for Slicer Assay

1. For the generation of template DNA 100 pmol of each oligonucleotide was incubated in a 40 μl reaction containing 1× Annealing buffer for 5 min at 95 °C followed by 45–60 min incubation at 37 °C. DNA templates were kept at −20 °C and used for several reactions.
2. In vitro transcription was performed in a 20 μl reaction containing 12.5 pmol template DNA, 1× transcription buffer, 1× $MgCl_2$-spermidin buffer, 2.5 mM NTP, 10 U T7 DNA polymerase, and 10 U RiboLock (Fermentas) for 1 h at 37 °C.
3. Reaction was stopped by DNase treatment (1 U) for 20 min at 37 °C.

4. RNA was extracted with 1 volume of Phenol–Chloroform mix (1:1), 1 volume of Chloroform, and precipitated with ethanol.
5. To verify the quality and quantity of in vitro generated RNAs, a small amount was 5′-end labeled and separated by denaturing Urea-polyacrylamide gel electrophoresis (PAGE). Labeled product at the same time provided a migration marker allowing the gel elution of unlabeled transcript.

3.2 Preparation of 5′-End Labeled RNA

1. RNA (0.25 % of the in vitro generated transcript) was first dephosphorylated with 0.25 U shrimp alkaline phosphatase (Roche) in a 10 μl reaction containing 1× reaction buffer (Roche) at 37 °C for 20 min.
2. After heat-inactivation at 75 °C for 5 min, the 5′-labeling occurred in a final volume of 15 μl for 15 min at 37 °C with 1× T4 PNK buffer (Fermentas), 1 pmol [γ-^{32}P] ATP (Perkin Elmer), and 5 U T4 PNK.
3. After the addition of 15 ml Formamid loading buffer 10 μl was separated by Urea-PAGE.
4. The kinase reaction was further used for the generation of radioactively labeled DNA probes employed in Northern Blot analysis of piRNAs. In a 50 μl reaction, 2.5–5 pmol of a DNA oligonucleotide (bearing perfect complementarity to the piRNA) was incubated with an equal amount of [γ-^{32}P] ATP and 10 U T4 Polynucleotide Kinase for 30 min at 37 °C. Reactions were purified in a Sephadex G25 spin column (Amersham).
5. For the labeling of the slicer substrate RNA, 0.5–1 pmol of gel-purified RNA was first dephosphorylated with shrimp alkaline phosphatase in a 15 μl reaction. After heat-inactivation 5′-labeling occurred in a 50 μl reaction for 15 min at 37 °C with 2.5 pmol [γ-^{32}P] ATP and 10 U T4 PNK. The reaction was purified with a Sephadex G25 column and adjusted with water to a concentration of 5 fmol RNA/μl.

3.3 Purification of Labeled RNA Substrates

1. The 42 nt in vitro transcribed substrate RNA was eluted from a 15 % Urea-polyacrylamide gel. For 25 ml of a 15 % gel mix 18.75 ml of Gel buffer I and 6.25 ml Gel buffer II. After the addition of 150 μl 10 % APS and 50 μl TEMED, the gel was poured.
2. After a pre-run of 20 min at 20 W in 1× TBE buffer, to warm up the gel, samples were loaded in the wells and separation occurred at 20 W for up to 1 h.
3. After the run, one glass plate was removed and the gel wrapped in transparency film. Small pieces of paper served as position markers, which were first socked in a diluted (1:10,000) [γ-^{32}P] ATP solution and after drying placed directly on the wrapped gel. A second sheet of transparency film fixed these radioactive-labeled position markers.

4. The gel was exposed to Storage PhosphorImager screens (GE Health). Screens were scanned in a Typhoon scanner and the image printed (without losing original size parameters) and placed beneath the gel, using the radioactive paper and their corresponding signals on the print-out as guides. RNA was cut out with a scalpel from the respective area of the gel.
5. Gel pieces were finely ground with a heat-rounded 1 ml pipette tip and combined with 400 μl 0.3 M NaCl. Samples were then placed in dry ice for complete freezing and were further incubated over night at 20 °C with constant and strong agitation.
6. To separate gel pieces from solution, samples were passed through centrifuge tube filters.
7. Samples were extracted with Phenol–Chloroform, Chloroform and precipitated with Ethanol in the presence of 10 μg Glycogen (Roche).
8. The RNA pellet was washed in 70 % ethanol, resuspended in water and the concentration determined.

3.4 Preparation Protein G Sepharose Beads Coupled to Antibodies

1. Protein G Sepharose beads were washed twice in wash buffer before the addition of antibody. Affinity-purified anti-Miwi rabbit polyclonal antibodies were bound to beads overnight on a rotating wheel at 4 °C. Approximately 0.15–0.5 μg purified antibody or 5–15 μl immune serum per 10 μl of protein G Sepharose slurry was used.
2. Unbound material was removed by two wash steps (each 1 ml) with wash buffer after the binding procedure.

3.5 Preparation of Testes Extract

1. The Piwi protein Miwi is abundantly detected in testes extracted from 20-day-old or adult mice. About half to one entire testes was shortly thawed (stored at −80 °C) in cold 1× PBS.
2. After removal of the Tunica albuginea (the layer of cells covering the seminiferous tubules within the testes), tissue was homogenized in 1 ml testes lysate buffer using a dounce homogenizer.
3. Extract was centrifuged twice at 9,200×g, 4 °C for 15 min. Supernatant was collected as testes lysate and stored on ice till further use.

3.6 Isolation of Endogenous Miwi Complexes from Mouse Testes

1. Testes extract was added to antibody-bound protein G Sepharose beads (prepared in Subheading 3.4). Incubation proceeded further at 4 °C under permanent agitation on a rotating wheel at slow speed. After incubation for 2–5 h, beads were washed 5 times with 1 ml wash buffer.
2. To reduce contamination from material sticking to the eppendorf tube used for lysate incubation, the suspended beads were

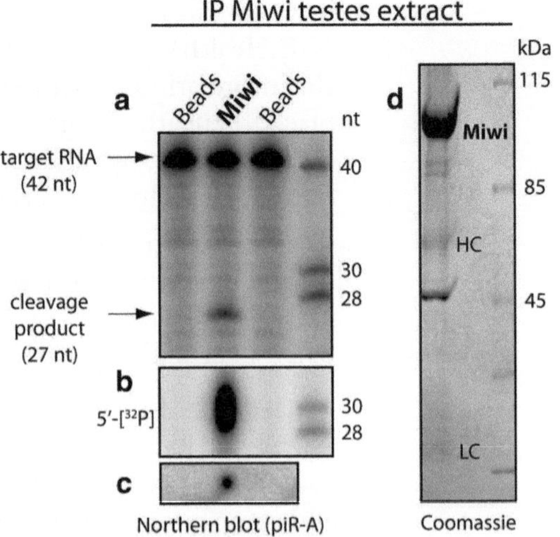

Fig. 2 Slicer Assay with Miwi complexes. (**a**) The 5′-end labeled 42 nt target RNA substrate bearing complete complementarity to the endogenous piRNA guide piRNA-A (piR-A) was incubated with immune purified Miwi complexes isolated from testis extract. Reactions were stopped after 120 min. The generation of a 27 nt cleavage product can be observed. The endonucleolytic cleavage is opposite to the 10th nucleotide of the guide (counted from the 5′ end) which is in agreement with reported activities of other Argonaute slicers. (**b**) The 5′-end labeling of piRNAs isolated from Miwi complexes. (**c**) Northern Blot of piRNA-A (piR-A). (**d**) Coomassie-stained SDS gel showing the separation of purified Miwi complexes from testis extract. HC: heavy chain, LC: light chain, RNA or protein marker bands are designated with the corresponding size in nt (nucleotides) and kDa (kilodalton), respectively

transferred to a new tube during the final wash step (by pouring or pipetting using a 1 ml tip with a cut end).

3. For slicer assays the suspended beads of the last wash step were divided as follows: 85 % was directly used for the cleavage assay. The residual 15 % was treated with proteinase K to isolate RNA for Northern Blot analysis or 5′ labeling of piRNA guides (Fig. 2).

3.7 Slicer Cleavage Assay

1. The immunopurified Miwi complexes bound to protein G Sepharose (don't exceed a volume of 10 μl slurry) were preincubated for at least 10 min in 10 μl Slicer assay buffer on ice. Beads were first covered with the buffer and then gently resuspended by tapping the tube.
2. The reaction was started by the addition of 10 μl of 5′-labeled substrate (10–50 fmol). RNA was denatured before at 55 °C for 5–10 min followed by the immediate transfer to ice.

3. The reaction was further performed in an Eppendorf thermoshaker for 10 min–2 h at 35 °C with constant changing modes of shaking (1,000 rpm, 10 s) and pause (90 s).
4. The reaction was stopped by a proteinase K treatment.
5. Samples were extracted with Phenol–Chloroform and precipitated with Ethanol in the presence of 10 μg Glycogen.
6. Reaction products were resolved by 15 % Urea-PAGE. Gels have been dried and exposed to storage PhosphorImager screens.

3.8 RNA Extraction and 5′-End Labeling to Detect Small RNAs in Miwi Complexes

1. After complete removal of the supernatant, washed beads were mixed with 300 μl proteinase K buffer and 10–15 μg proteinase K. Samples were incubated at 42 °C for 20 min with constant agitation.
2. After one extraction step with Phenol–Chloroform, RNA was precipitated with Ethanol in the presence of 10 μg Glycogen.
3. RNAs were 5′-end labeled as described above.

3.9 Northern Blotting to Detect the Presence of Specific Small RNA Guide in Miwi Complexes

1. To detect specific piRNA species, RNA samples were separated by 15 % Urea–PAGE and transferred overnight on to Nylon membranes (Amersham) with 0.5× TBE buffer using a semidry electro blot machine (5 V, overnight).
2. After UV-crosslinking ("Stratalinker" auto cross-link mode) the membranes were rinsed in 2× SSC and dried for storage in between two sheets of Whatman paper. Alternatively, they were immediately pre-hybridized in Church buffer for at least 2 h at 37 °C (for small RNA detection).
3. Depending on the number of hybridization experiments and overall probe concentration, about 50–100 % of the probe (Subheading 3.1, **step 2**) was denatured at 95 °C for 5 min followed by an immediate transfer to ice and added directly in Church buffer. Dependent on the size of the membrane 5–10 ml of hybridization buffer was used resulting in a probe dilution of 1:100–1:600. Hybridization occurred overnight at 37 °C.
4. Membranes were washed twice for 15 min, first in 0.2× SSC, 0.1 % SDS and then in 0.1× SSC, 0.1 % SDS at the hybridization temperature. After washes, the membrane was wrapped in transparency film and exposed to a PhosphorImager screen and scanned.

Acknowledgments

We gratefully acknowledge help and advice from Pietro Spinelli, Zhaolin Yang, Elisa Cora, Bruno Rodino, and Sara Silva in developing the assay protocol. We thank Kuan-Ming Chen, Radha

Raman Pandey, and David Homolka for their critical comments on the manuscript. Help from Magdalena Wojtas for preparation of the figures is acknowledged. Work in the Pillai lab is supported by a European Research Council Starting Grant (pisilence) from the European Union to R.S.P.

References

1. Ghildiyal M, Zamore PD (2009) Small silencing RNAs: an expanding universe. Nat Rev Genet 10(2):94–108
2. Siomi MC, Sato K, Pezic D, Aravin AA (2011) PIWI-interacting small RNAs: the vanguard of genome defence. Nat Rev Mol Cell Biol 12(4):246–258
3. Malone CD, Hannon GJ (2009) Small RNAs as guardians of the genome. Cell 136(4):656–668
4. Martinez J, Patkaniowska A, Urlaub H, Luhrmann R, Tuschl T (2002) Single-stranded antisense siRNAs guide target RNA cleavage in RNAi. Cell 110(5):563–574
5. Carmell MA, Xuan Z, Zhang MQ, Hannon GJ (2002) The Argonaute family: tentacles that reach into RNAi, developmental control, stem cell maintenance, and tumorigenesis. Genes Dev 16(21):2733–2742
6. Sashital DG, Doudna JA (2010) Structural insights into RNA interference. Curr Opin Struct Biol 20(1):90–97
7. Pak J, Fire A (2007) Distinct populations of primary and secondary effectors during RNAi in C. elegans. Science 315(5809):241–244
8. Horwich MD et al (2007) The Drosophila RNA methyltransferase, DmHen1, modifies germline piRNAs and single-stranded siRNAs in RISC. Curr Biol 17(14):1265–1272
9. Kirino Y, Mourelatos Z (2007) The mouse homolog of HEN1 is a potential methylase for Piwi-interacting RNAs. RNA 13(9):1397–1401
10. Saito K et al (2007) Pimet, the Drosophila homolog of HEN1, mediates 2′-O-methylation of Piwi- interacting RNAs at their 3′ ends. Genes Dev 21(13):1603–1608
11. Yu B et al (2005) Methylation as a crucial step in plant microRNA biogenesis. Science 307(5711):932–935
12. Ma JB, Ye K, Patel DJ (2004) Structural basis for overhang-specific small interfering RNA recognition by the PAZ domain. Nature 429(6989):318–322
13. Ma JB et al (2005) Structural basis for 5′-end-specific recognition of guide RNA by the A. fulgidus Piwi protein. Nature 434(7033):666–670
14. Tian Y, Simanshu DK, Ma JB, Patel DJ (2011) Structural basis for piRNA 2′-O-methylated 3′-end recognition by Piwi PAZ (Piwi/Argonaute/Zwille) domains. Proc Natl Acad Sci U S A 108(3):903–910
15. Simon B et al (2011) Recognition of 2′-O-methylated 3′-end of piRNA by the PAZ domain of a Piwi protein. Structure 19(2):172–180
16. Frank F, Sonenberg N, Nagar B (2010) Structural basis for 5′-nucleotide base-specific recognition of guide RNA by human AGO2. Nature 465(7299):818–822
17. Wang Y et al (2008) Structure of an argonaute silencing complex with a seed-containing guide DNA and target RNA duplex. Nature 456(7224):921–926
18. Wang Y et al (2009) Nucleation, propagation and cleavage of target RNAs in Ago silencing complexes. Nature 461(7265):754–761
19. Filipowicz W, Bhattacharyya SN, Sonenberg N (2008) Mechanisms of post-transcriptional regulation by microRNAs: are the answers in sight? Nat Rev Genet 9(2):102–114
20. Buhler M, Haas W, Gygi SP, Moazed D (2007) RNAi-dependent and -independent RNA turnover mechanisms contribute to heterochromatic gene silencing. Cell 129(4):707–721
21. Buhler M, Moazed D (2007) Transcription and RNAi in heterochromatic gene silencing. Nat Struct Mol Biol 14(11):1041–1048
22. Buhler M, Spies N, Bartel DP, Moazed D (2008) TRAMP-mediated RNA surveillance prevents spurious entry of RNAs into the Schizosaccharomyces pombe siRNA pathway. Nat Struct Mol Biol 15(10):1015–1023
23. Buhler M, Verdel A, Moazed D (2006) Tethering RITS to a nascent transcript initiates RNAi- and heterochromatin-dependent gene silencing. Cell 125(5):873–886
24. Liu J et al (2004) Argonaute2 is the catalytic engine of mammalian RNAi. Science 305(5689):1437–1441
25. Rivas FV et al (2005) Purified Argonaute2 and an siRNA form recombinant human RISC. Nat Struct Mol Biol 12(4):340–349
26. Song JJ et al (2003) The crystal structure of the Argonaute2 PAZ domain reveals an RNA binding motif in RNAi effector complexes. Nat Struct Biol 10(12):1026–1032
27. Song JJ, Smith SK, Hannon GJ, Joshua-Tor L (2004) Crystal structure of Argonaute and its

implications for RISC slicer activity. Science 305(5689):1434–1437
28. Meister G et al (2004) Human Argonaute2 mediates RNA cleavage targeted by miRNAs and siRNAs. Mol Cell 15(2):185–197
29. Yuan YR et al (2005) Crystal structure of *A. aeolicus* argonaute, a site-specific DNA-guided endoribonuclease, provides insights into RISC-mediated mRNA cleavage. Mol Cell 19(3): 405–419
30. Martinez J, Tuschl T (2004) RISC is a 5′ phosphomonoester-producing RNA endonuclease. Genes Dev 18(9):975–980
31. Orban TI, Izaurralde E (2005) Decay of mRNAs targeted by RISC requires XRN1, the Ski complex, and the exosome. RNA 11(4):459–469
32. Saito K et al (2006) Specific association of Piwi with rasiRNAs derived from retrotransposon and heterochromatic regions in the Drosophila genome. Genes Dev 20(16):2214–2222
33. Gunawardane LS et al (2007) A slicer-mediated mechanism for repeat-associated siRNA 5′ end formation in Drosophila. Science 315(5818): 1587–1590
34. Brennecke J et al (2007) Discrete small RNA-generating loci as master regulators of transposon activity in Drosophila. Cell 128(6): 1089–1103
35. Reuter M et al (2011) Miwi catalysis is required for piRNA amplification-independent LINE1 transposon silencing. Nature 480:264–267. doi:10.1038/nature10672
36. De Fazio S et al (2011) The endonuclease activity of Mili fuels piRNA amplification that silences LINE1 elements. Nature 480: 259–263
37. Cifuentes D et al (2010) A novel miRNA processing pathway independent of Dicer requires Argonaute2 catalytic activity. Science 328(5986): 1694–1698
38. Cheloufi S, Dos Santos CO, Chong MM, Hannon GJ (2010) A dicer-independent miRNA biogenesis pathway that requires Ago catalysis. Nature 465(7298):584–589
39. Miyoshi K, Tsukumo H, Nagami T, Siomi H, Siomi MC (2005) Slicer function of Drosophila Argonautes and its involvement in RISC formation. Genes Dev 19(23):2837–2848
40. Irvine DV et al (2006) Argonaute slicing is required for heterochromatic silencing and spreading. Science 313(5790):1134–1137
41. Addo-Quaye C, Eshoo TW, Bartel DP, Axtell MJ (2008) Endogenous siRNA and miRNA targets identified by sequencing of the Arabidopsis degradome. Curr Biol 18(10): 758–762
42. Darricarrere N, Liu N, Watanabe T, Lin H (2013) Function of Piwi, a nuclear Piwi/Argonaute protein, is independent of its slicer activity. Proc Natl Acad Sci U S A 110(4): 1297–1302
43. Bagijn MP et al (2012) Function, targets, and evolution of *Caenorhabditis elegans* piRNAs. Science 337(6094):574–578

Chapter 10

Small RNA Library Construction from Minute Biological Samples

Jessica A. Matts, Yuliya Sytnikova, Gung-wei Chirn, Gabor L. Igloi, and Nelson C. Lau

Abstract

Increasingly, the discovery and characterization of small regulatory RNAs from a variety of organisms have all required deep-sequencing methodologies. However, the crux to successful deep-sequencing analysis depends upon optimal construction of a cDNA library compatible for the high-throughput sequencing platform. Challenges to small RNA library constructions arise when dealing with minute tissue samples because certain structural RNA fragments can dominate and mask the desired characterization of regulatory small RNAs like microRNAs (miRNAs), endogenous small interfering RNAs (endo-siRNAs), and Piwi-interacting RNAs (piRNAs). Here, we describe methods that improve the chances of constructing a successful library from small RNAs isolated from minute tissues such as enriched follicle cells from the *Drosophila* ovarium. Because the ribosomal RNA (rRNA) fragments are frequently the major contaminants in small RNA preparations from minute amounts of tissue, we demonstrate the utility of antisense oligonucleotide depletion and an acryloylaminophenylboronic acid (APB) polyacrylamide gel system for separating the abundant 2S rRNA in *Drosophila* from endo-siRNAs and piRNAs. Finally, our methodology generates libraries amenable to multiplex sequencing on the Illumina Hi-Seq platform.

Key words Small RNAs, Illumina deep-sequencing, Library construction

1 Introduction

Among the various methods to detect small endogenous RNAs in organism samples, the procedure to convert RNAs into a library of cDNAs followed by sequencing is considered to be the most comprehensive technique because it enables discovery of the molecules' sequence as well as confirmation of its identity. For example, the seminal discoveries that expanded the catalog of microRNAs beyond lin-4 and let-7 resulted from the sequencing of cDNA libraries from small RNAs purified from nematode, fly and human cells [1–3]. With the commercialization and adoption of high-throughput next generation

Jessica A. Matts and Yuliya Sytnikova have equally contributed to this work

Mikiko C. Siomi (ed.), *PIWI-Interacting RNAs: Methods and Protocols*, Methods in Molecular Biology, vol. 1093,
DOI 10.1007/978-1-62703-694-8_10, © Springer Science+Business Media, LLC 2014

sequencing technologies such as 454, Illumina, and ABI SoLID, the detection of molecules based on the number of reads have lent to unparalleled sensitivity and accuracy in small RNA quantitation.

Despite the immense depth of sequencing now achievable by high-throughput systems like the Illumina Hi-Seq 2000 capable of routinely generating >200 million reads per flow cell lane, the success of a small RNA sequencing run depends most on the composition of the library in terms of being free of undesired degradation remnants or naturally abundant short ribosomal RNA fragments. For example, *Drosophila* and other insects generate an abundant 31-nucleotide (nt) 2S rRNA that migrates very closely to the piRNAs that range from 24 to 32 nt long [4]. In addition, *Xenopus* eggs also contain a slew of ribosomal RNA fragments that range from 23 to 35 nt long, which also co-migrate in size with *Xenopus* piRNAs and miRNAs [5]. When there are plenty of tissues and cells from which to generate a cell lysate, this lysate can be subjected to immunoprecipitation (IP) of ARGONAUTE and PIWI proteins, or with cation exchange chromatography [6], both of which can quite effectively deplete undesired contaminating RNAs.

However, when one wishes to profile small RNAs from minute samples, such as a single egg or from a very small population of cells enriched in a particular cell type, the IP and chromatography methods are not practical, and typically just total RNA is isolated. From these total RNA preparations, rRNA fragments persist and will become a major nuisance. For example, when total small RNAs from single *Xenopus* eggs were profiled, the rRNA fragments reduced the representation of other small regulatory RNAs down to below 20 % of the library [5]. This issue can be partially mitigated by sequencing libraries on the Illumina Hi-Seq platform versus the Illumina Genome Analyzer (GA) platform because the ~20-fold increase in depth from the former platform may yield enough desired small regulatory RNA reads despite sacrificing the non-useful rRNA contaminants. However, new considerations in the multiplexing of small RNA libraries must be followed for the Illumina Hi-Seq platform because the lower stringency of base-calling versus the GA platform can also reduce the yield of reads passing quality. In the first set of methods detailed in this chapter, we describe our experiences in isolating a small sample of tissues, an enriched population of follicle cells from the *Drosophila* ovarium, and our considerations in generating small RNA libraries from these minute samples for the Illumina Hi-Seq platform, which includes a new format of linkers that are amenable for multiplex sequencing of small RNA libraries.

In the second set of methods, we describe the adaptation of a boronate affinity gel matrix applied to the resolution of small amounts of RNA from *Drosophila* ovary cells. The boronate gel matrix consists of a denaturing polyacrylamide gel impregnated with acryloylaminophenylboronic acid (APB), and short RNAs (<~80 nt) with unmodified 2′–3′ cis-diols will exhibit a stronger dynamic affinity to the boronate than endo-siRNAs and piRNAs which are naturally methylated at the 3′ end by Hen-1 on the 2′ OH [7, 8]. With standard polyacrylamide electrophoresis, the abundant 2S rRNA and other rRNA fragments can co-migrate or resolve poorly from piRNAs and endo-siRNAs. However, on an APB gel the rRNA fragments are retarded while bonafide piRNAs and endo-siRNAs migrate faster, thus facilitating further the removal of the contaminating RNAs from the regulatory small RNAs.

As biologists begin to interrogate the small RNA profiles of particular niches of cells, from stem cells to specialized neuronal cell types, the need to improve methodologies to generate libraries from minute samples will become more evident. These procedures we have developed will increase the likelihood that properly diverse cDNA libraries can be constructed, and although our reagents are based on lab-made stocks, the antisense oligo-mediated depletion step and the boronate affinity gel matrix can be applied to steps from commercial small RNA library construction protocols.

2 Materials

2.1 *Drosophila* Culture and Ovary Dissection

1. Standard fly food in bottles with enough extra yeast added to make a fine layer on the top of the food.
2. Flynap or carbon dioxide venting flypad for anesthetizing flies.
3. Stereo dissection microscope.
4. Paintbrush for fly pushing.
5. Fine pointed Dumont 50 tweezers.
6. Watch glass dish chilled in a small box of ice.
7. Chilled 1× PBS.

2.2 Follicle Cell Enrichment

1. Trypsin (Sigma).
2. 4′,6-diamidino-2-phenylindole (DAPI) stain and Vectashield (Vector labs).
3. 1 mg/ml Concanavalin A solution (Sigma).
4. 1 % Formaldehyde in 1× PBS (for fixing follicle cells).
5. 3.7 % Formaldehyde in 1× PBS (for fixing ovaries).
6. Programmable tube mixer (i.e. Thermomixer, Eppendorf).

2.3 Small RNA Purification and Antisense Oligo Depletion of 2S rRNA

1. TriReagent or TriZol solutions (Molecular Research Center or Invitrogen).
2. 2× RNA Urea-TE loading buffer: 8 M Urea, 100 mM Tris–HCl (pH 8.0), 50 mM EDTA, 0.05 % w/v bromophenol blue, 0.05 % w/v xylene cyanol FF.
3. 10 bp ladder (Invitrogen).
4. Gamma-^{32}P-labeled markers (18mer and 34mer).

5. SyberGreen II RNA stain (Invitrogen).
6. Phosphorimaging cassette and phosphorimager (i.e., Typhoon Scanner, GE Healthcare).
7. Siliconized 1.7-mL microcentrifuge tubes.
8. 0.3 M NaCl.
9. 10 μM biotinylated DNA oligo antisense to 2S rRNA: TACA ACCCTCAACCATATGTAGTCCAAGCATACAACCCTCA ACCATATGTAGTCCAAGCAGTCGA-3′biotin.
10. 0.5 M EDTA (pH8.0).
11. 20× SSC: 3 M NaCl, 0.3 M sodium citrate (pH 7.0).
12. 0.5 M NaCl.
13. Glycogen from blue mussel, 20 mg/mL (Sigma).

2.4 Acryloylamino-phenylboronic Acid Polyacrylamide Gel

1. Acryloylaminophenylboronic acid (APB, tRNA Probes, Inc.) [originally produced by the protocol of Ref. 9].
2. Sequagel Urea gel system (National Diagnostics).
3. 50× Tris-Acetate–EDTA (TAE) buffer: 2 M Tris-Acetate, 50 mM EDTA (pH 8.0).
4. Vertical electrophoresis system (i.e., Model V16 from Whatman, Inc.).
5. 2× Urea-TE loading buffer: 8 M Urea, 10 mM Tris–HCl (pH 8.0), 50 mM EDTA, 0.05 % w/v bromophenol blue, 0.05 % w/v xylene cyanol FF.
6. Synthetic RNA Marker oligonucleotides of any sequence at 18, 22, and 32 nt long.
7. Synthetic piRNA with standard terminal 2′-OH: AGGAA AGUUGUGCACACUUGUAAUCCGAAA.
8. Synthetic piRNA with terminal 2′-*O*-Me: UGGGAUUACA AGUGUGCACAACUUUCCUGmC.

2.5 Linker-Ligation Small RNA Library Construction from Minute Samples

1. Chimeric marker oligonucleotides where uppercase bold represents an RNA base, lowercase italics is a standard DNA base, and "dU" is deoxy-uridine, which can be degraded by uracil-deoxyglycosylase (UDG):

   34 nt chimeric marker: **CAGUAC** *ggatcca* dUdUdUdU *tat-gctc* **AGCGUACGAA**

   18 nt chimeric marker: **C** agtac dUdUdU gctag **CUAA.**
2. 3′ Linker adaptor: p-*cgtcgtatgccgtcttctgcttgt*-/3AmMO/, where /3AmMO/ is a 3′ amino modifier and is chemical phosphorylated on the 5′ end.
3. Original 5′ Adaptor: **GUUCAGAGUUCUACAGUCCG-ACGAUC**.

4. 5′ Hi-Seq adaptors with barcodes underlined, RNA bases in uppercase bold, DNA bases in lowercase italics, and "N" represents random incorporation of all four DNA bases:
   Barcode CAA: gttcagagttctacagtccgacgatc *NNN* CAA**AA**
   Barcode ACC: gttcagagttctacagtccgacgatc *NNN* ACC**AA**
   Barcode GUU: gttcagagttctacagtccgacgatc *NNN* **GUUAA**
   Barcode UGG: gttcagagttctacagtccgacgatc *NNN* **UGGAA**.
5. Reverse Transcription primer & 5′ PCR primer: caagcagaagacggcata.
6. 3′ PCR primer: aatgatacggcgaccaccgacaggttcagagttctacagtccga.
7. Mth RNA ligase (New England Biolabs).
8. Polynucleotide kinase (PNK) (New England Biolabs).
9. Uracil-DNA glycosylase (New England Biolabs).
10. Phusion Polymerase (New England Biolabs).
11. T4 RNA ligase I enzymes (New England Biolabs).
12. RiboLock RNase inhibitor, 40 U/μl (Thermo Scientific).
13. 5× RNA ligase buffer: 250 mM HEPES (pH 8.3), 50 mM $MgCl_2$, 16.5 mM DTT, 50 μg/mL BSA, 41.5 % glycerol.
14. 10 mM ATP.
15. Low melting temperature agarose (i.e., Agarose II, Amresco).
16. Gel extraction kit (Qiagen).
17. RNA Clean & Concentrator kit (Zymo Research).
18. SuperScript III Reverse Transcriptase (Invitrogen).
19. Quant-iT Pico-green dsDNA assay kit (Invitrogen).
20. Zero-Blunt TOPO PCR cloning kit for sequencing (Invitrogen).

3 Methods

3.1 Harvesting Ovaries and Enriching for Follicle Cells

About 3 days prior to dissections, flies should be given extra yeast in the food in order to fatten up the ovaries so that they are easily distinguishable and removable from the rest of the abdomen (*see* **Note 1**). Typically, ~400 female flies are dissected in one sitting to obtain ~100 μl of tissue. Follicle cell enriching must be performed immediately after one dissection sitting, and practice is required to enable obtaining tissue efficiently. The amount of trypsin to add varies depending on amount of tissue (usually 7.5 μl of a 10× trypsin stock from Sigma per 100 μl of tissue yields a good enriched follicle cell sample).

1. Anesthetize fattened flies with either flynap or carbon dioxide, and sort for females (larger yet lighter abdomen).
2. Transfer females to a watch glass containing cold 1× PBS. Use fine tweezers to tease out the ovaries by pulling away from the abdomen and place ovaries in a microcentrifuge tube containing ice-cold 1× PBS.

3. Wash ovaries once in 1 mL cold 1× PBS (save a 20 μl drop for DAPI staining).
4. Add 700 μl of cold 1× PBS to the ovaries, then add 77 μl of a 10× trypsin stock. Vortex briefly to mix.
5. Place tube in a thermoshaker to shake at 1,400 rpm (or the maximum rpm) for 20 min at 30 °C. Shaking must be vigorous (>1,200 rpm) to ensure that the majority of the nurse cells and oocytes are properly lysed during enzymatic digestion.
6. Rest the tube on ice to allow undigested tissue to settle and remove the supernatant of liberated follicle cells to a new tube which is centrifuged at 4 °C for 8 min at 2,000 × *g* to pellet the follicle cells.
7. Repeat **steps 4–6** at least two more times on remaining undigested tissues to collect additional enriched follicle cell samples. The final remaining tissue mostly consists of remaining mostly consists of fertilized eggs (up to 70 %) whose egg shells are resistant to trypsin digestion and can be discarded.
8. Wash and centrifuge the follicle cell pellet once in ice-cold 1× PBS (2 min, 1,600 × *g*), and resuspend pellet in 300 μl of cold 1× PBS. Remove 5 μl for DAPI staining verification. Spin down cells, remove PBS, and store the cell pellet at −80 °C until ready to start the small RNA library construction.
9. To verify efficacy of follicle cell enrichment, drops of dissected ovaries and follicle cells are placed on a microscope glass slide previously coated with Concanavalin A, which promotes cell adhesion to glass. After 5 min at room temperature, cells are fixed with formaldehyde in PBS for 5 min, then dipped in a 50 mL solution of 10 μg/ml DAPI in 1× PBS in a coplin jar for 5 min. Cells and ovaries are then mounted in Vectashield with a cover slip and observed under a fluorescent microscope with a DAPI filter to visualize the clear difference in size and morphology between intact ovaries and follicle cells (Fig. 1c).

3.2 Small RNA Purification and Antisense Oligo-Mediated 2S rRNA Depletion

Typically, two rounds of dissections and follicle cell enrichment would yield 10 μg of total RNA. However, we have been able to perform library construction from as little as 1–3 μg of total RNA [5]. Total RNA is extracted using TriReagent (also known as Trizol), and the very small pellet is resuspended in 20 μl of water.

1. Small RNAs are purified by gel purification from a 15 % urea denaturing gel by electrophoresis at 30 W for 75 min or until the darker Bromphenol blue dye is near the bottom of the gel. Total RNA is first denatured in 2× Urea Loading buffer with heating at 95° for 5 min, and samples are run with a 10 bp ladder and/or radioactive RNA markers.
2. If using radioactive RNA markers, expose gel covered in plastic wrap against a phosphor plate for 15 min. If only a 10 bp DNA

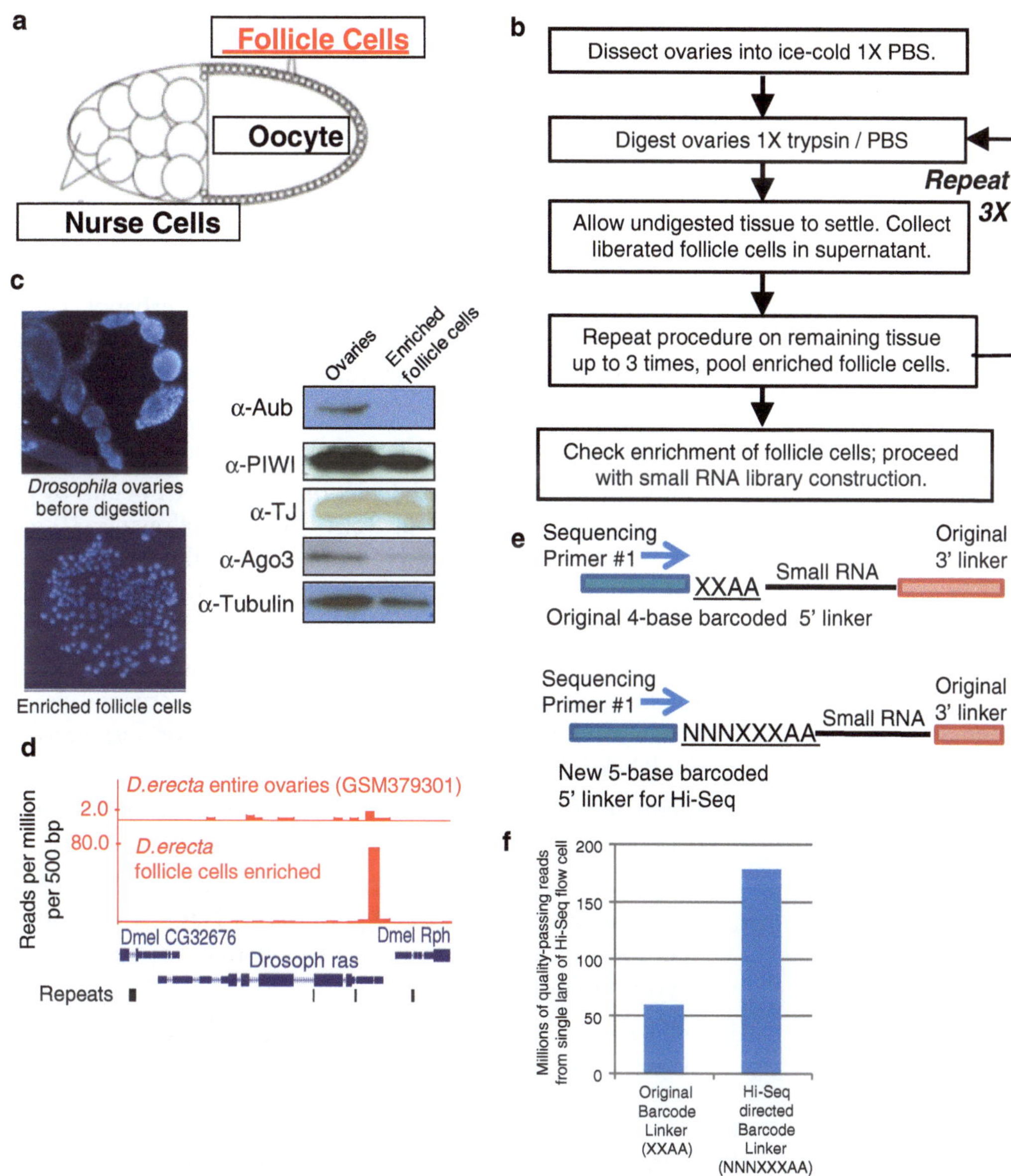

Fig. 1 *Drosophila* ovary follicle cell enrichment and small RNA library construction from this minute tissue sample. (**a**) A schematic of a *Drosophila* egg chamber highlighting the relative small size of the follicle cells. (**b**) Flowchart of the procedure to enrich follicle cells from *Drosophila* ovaries. (**c**) Validation of follicle cell enrichment procedure with 4′,6-diamidino-2-phenylindole (DAPI) staining in left panels and western blots that show retention of Piwi and Traffic Jam (TJ) proteins concomitant with depletion of Aubergine (Aub) and Argonaute-3 (Ago3) proteins. (**d**) Counts of genic 3′UTR-directed piRNAs from the *Raspberry* (*ras*) gene are enriched from the follicle cells compared to sequencing from total ovary RNA. (**e**) Schematic comparing the original 5′ linker barcode format that was suitable for sequencing on the Illumina Genome Analyzer and the new barcode format we developed for the Illumina Hi-Seq platform. (**f**) A small RNA library with the original 4-base barcode suffers from poor read quality discrimination, while the simple addition of a 3-base random element in the new 5-base barcode greatly increases the number of quality-passing reads

ladder is used, stain the gel with SyberGreen II for 10 min. Phosphor plate or gel is directly scanned on the image scanner (phosphorimaging or fluorescent setting). Print a full-scale picture of the gel, and placing this behind the gel, cut out the region of small RNAs from ~20 to 30 nt (the 2S rRNA migrates at 31 nt).

3. Elute the gel samples overnight at 4 degrees C in 500 μl of 0.3 M NaCl.
4. Precipitate the RNA in 2 volumes of absolute ethanol. 1 μg of mussel glycogen greatly enhances precipitation of small RNAs, while using siliconized tubes reduces nonspecific adherence of RNA to plastic.
5. Resuspend the RNA in 25 μl of dH_2O.
6. Antisense oligo-mediated depletion of 2S rRNA begins by prebinding oligo to the 25 μl RNA with the following: 10 μl of anti-2S rRNA biotin oligo (10 μM), 2 μl of 0.5 M EDTA (pH 8.0), 10 μl of 20× SSC, and 68 μl of dH_2O. Heat the sample to 70 °C for 5 min and slowly cool to 37 °C.
7. Meanwhile, prepare 600 μl per sample of MagneSphere streptavidin beads by washing according to manufacturer's instruction with 0.5× SSC, and add the 100 μl beads suspension to the RNA/oligo mix.
8. Incubate at room temperature for 20 min with mild agitation (can be placed on a rotisserie rocker to rotate).
9. Bind beads to a magnetic stand for 3 min, remove the supernatant to a new tube, add 200 μl of 0.5 M NaCl, 1 μl of glycogen, and 1 ml of absolute ethanol to supernatant and precipitate overnight at −20° C.
10. Next day pellet RNA and resuspend in 10 μl of dH_2O for library construction.

3.3 Boronate Affinity Gel Electrophoresis to Resolve piRNAs from Other Small RNAs

Another way to purify piRNAs from 2S rRNA and other abundant rRNA fragments is to exploit the differential affinity of modified versus unmodified terminal 3′-ends of nucleic acids to special gel matrixes during electrophoresis. For example, an acrylamide gel matrix impregnated with a boronate group can react in a dynamic equilibrium with the 3′-terminal 2′ and 3′ hydroxyl groups of unmodified short RNAs (Fig. 2a), but this reaction does not occur when there is a 2′-*O*-methyl group. Boronate groups are incorporated by adding APB to a standard denature urea polyacrylamide gel [9]. Short RNAs of similar lengths but with 3′-terminal modifications will not sufficiently resolve simply by size on a standard polyacrylamide gel (Fig. 2b, c), however when APB is added to 10 % of the acrylamide concentration (e.g., for a 20 % polyacrylamide gel we add APB to 2 %), the piRNAs with a 3′-terminal 2′-*O*-methyl group now migrate much faster in the gel and separate very effectively from unmodified rRNAs. Electrophoresis with

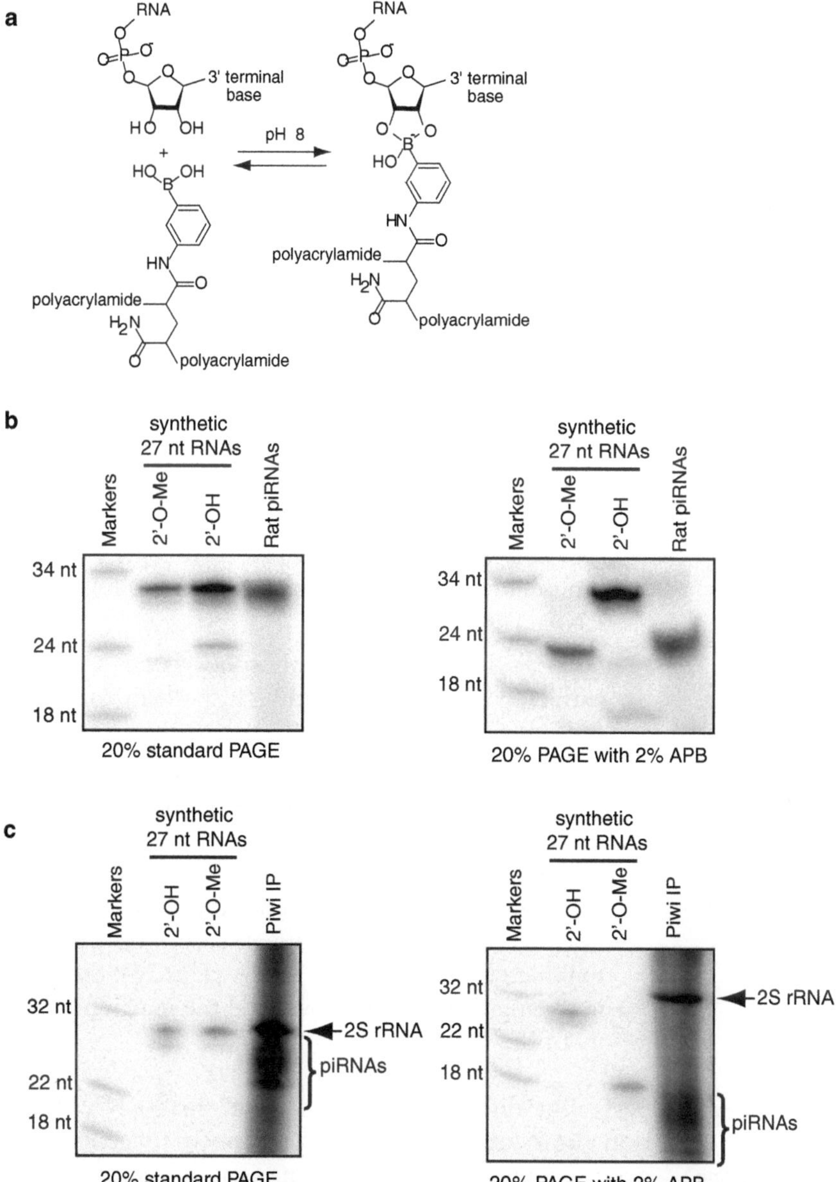

Fig. 2 Boronate affinity gel electrophoresis of small regulatory RNAs. (**a**) Scheme describing the structure of acryloylaminophenylboronic acid (APB) linked to polyacrylamide, and the reversible, dynamic reaction between boronate and the 3′ end of unmodified RNAs such as rRNA fragments. Small regulatory RNAs such as endogenous siRNAs and Piwi-interacting RNAs are methylated on the 3′-terminal 2′-OH, which prevents them from reacting with boronate. (**b**) 5′ radiolabeled synthetic RNAs with a 3′-terminal 2′-OH or 2′-*O*-Me electrophoresed along with radiolabeled rat piRNAs on a standard and APB polyacrylamide gel. The methylated synthetic RNA and rat piRNAs migrate much faster than their typical size when compared to the unmodified RNAs. (**c**) The mobility of *Drosophila* piRNAs are increased in the APB gel, allowing for better resolution of the piRNAs from the contaminating and highly abundant 2S rRNA that co-precipitate nonspecifically in a Piwi immunoprecipitation (IP). The *Drosophila* piRNAs from OSS cells were derived from a Piwi antibody IP performed with protein A/G magnetic beads followed by RNA extraction with TRI Reagent RT

APB gels must utilize Tris-Acetate (TAE) buffer instead of the typical Tris-Borate (TBE) buffer used for polyacrylamide gels because excess borate ions perturb complex formation between the boronate and the RNA.

1. Prepare a 20 % APB-polyacrylamide mix for a 30 ml, 0.8 mm thick gel with 24 ml of Sequagel Concentrate, 3 ml of Sequagel Diluent, 1.5 ml of a 50× TAE stock, 600 mg of APB powder, and 1 ml of dH_2O.
2. Warm the mix to 60 °C till APB powder dissolves completely and then cool back to room temperature.
3. Add 120 μl of 10 % APS and 12 μl of TEMED, mix and pour into the glass plates and spacers, and let the gel polymerize for at least 1 h at room temperature.
4. In our control experiments which are optional for actual library construction, we electrophoresed the same RNA samples on a standard Urea-TAE gel versus an APB gel to demonstrate the different mobility behaviors of piRNAs relative to synthetic RNA markers.
5. Natural RNAs were dephosphorylated with phosphatase, extracted with phenol and chloroform, and then ethanol precipitated.
6. Natural and synthetic RNAs are then labeled with ^{32}P-γ-ATP and PNK and denatured in Urea Loading Dye for 5 min at 95 °C prior to gel loading. Note in our demonstration that on a standard polyacrylamide gel, residual 2S rRNAs persisting in a Piwi immunoprecipitation do not resolve well from piRNAs, however on an APB gel, piRNAs can now be resolved several millimeters away from 2S rRNA (Fig. 2c). *See* **Note 2** regarding being careful not to overload the maximum binding capacity of the APB gel.

3.4 Generating Small RNA Libraries from Minute RNA Quantities for Multiplexing on the Illumina Hi-Seq Platform

Several modifications of our original small RNA library construction protocol [1] have been described to generate libraries for deep-sequencing on the Illumina Genome Analyzer [4, 5, 10–18]. We further extend the list of modifications with a protocol geared towards preparations for the Illumina Hi-Seq sequencer and starting with a small quantity of RNA. A different format for barcoding and multiplexing in the 5′ linker adapter is required because the Hi-Seq machine's colony-calling algorithm now demands significant base diversity in the first 4 bases sequenced, as evidenced by our previous 4-base barcode immediately downstream sequencing primer #1 yield 1/5th of the theoretical maximum number of reads (~60 million versus ~300 million, Fig. 1e). The introduction of three random nucleotides before a 5 bp barcode sequence creates sufficient color diversity for the Hi-Seq colony-calling algorithm to approach sufficient number of quality-passing reads

(Fig. 1e, f). *See* **Note 3** on additional considerations that motivated the modification described here. The steps below highlight additional specific modifications to include in other previously described small RNA library construction procedures.

1. Previously we relied upon chemical adenylation as the most cost-effective method to generate pre-adenylated 3′ Linker adaptors, but the organic chemistry steps are not routine procedures for molecular biologists. We now employ enzymatic adenylation of the 3′ Linker adaptor with the Mth RNA ligase in the 5′-Adenylation Kit sold by New England Biolabs, which nearly quantitatively adds an adenylated moiety to the 5′ end of a monophosphorylated oligonucleotide.
2. Although gel-purified RNAs from passive elution in NaCl is typically concentrated by ethanol precipitation with glycogen carrier in siliconized tubes, for very small samples we have observed decreased loss of RNA by using the RNA Clean & Concentrator kit from Zymo Research, following the manufacturer's protocol for total RNA (>17 nt). We also recommend increasing the volume of ethanol 3× for the mix in the first two steps of the standard manufacturer's protocol. Elutions are performed with 20 μl of dH_2O.
3. The 5′ Linker adaptor ligation reaction components and conditions remain the same as previous protocols except the use of the new barcoded 5′ Linker adaptors for the Hi-Seq sequencer. When one has included the new chimeric RNA/DNA markers of 18 and 34 nt length with internal deoxy-uracils as spiked-in radiolabeled markers to facilitate visualizing the shift of ligated products via gel-electrophorese and phosphorimaging, there is the optional post-reaction step of adding 5 units of uracil DNA glycosylase to the 5′ Linker-ligation reaction and incubating for 20 min at 37 °C.
4. Reverse transcription (RT) with Superscript III reverse transcriptase and PCR with Phusion polymerase can be directly performed from ¼ or ½ of the 5′ Linker-ligation reaction without the need of purifying the ligated molecules from unligated molecules, following the same protocol as previous methods [4, 5, 15–18]. However, a common problem in amplifying a library that began from minute starting RNA is that linker-linker dimers can dominate in the final PCR, regardless of whether one performs or skips the gel purification of the ligated products after the 5′ Linker adapter ligation reaction. By reducing the number of PCR cycles (no more than 15 cycles) in the initial amplification rounds from the RT reaction, this PCR sample can be gel purified on a 4 % low melting temperature agarose gel for the desired 95–115 nt long amplicons from the linker-linker dimer that is ~80 nt long.

5. From this purification, an additional 15 cycles of PCR can be applied to generate enough material for small scale cloning with a standard PCR cloning vector like in the Zero-Blunt TOPO PCR cloning kit followed by Sanger sequencing verification. Purified PCR-amplified DNA can be suitably quantitated for concentration with the Quant-iT Pico-green dsDNA assay kit. This library sample is thus ready for deep sequencing onto the Hi-Seq sequencer.

4 Notes

1. Female flies that have been fattened up with yeast prior to ovary dissection yield larger egg chambers and facilitate the manual dissection of ovaries, which in turn enhances the harvest of follicle cells. Our enrichment procedure significantly depletes the oocytes and nurse cells in the *Drosophila* egg chamber, presumably from loss of structural integrity after trypsinization, however all cell-cycle stages of follicle cells are isolated in our procedure. For more extensive purifications of specific staged follicle cells, *see* refs. 19, 20.
2. The boronate affinity gel was first described as a means to resolve and detect modifications on transfer RNAs [9]. These authors noted the limitation of the boronate gel in terms of overloading the capacity of the gel and the maximum resolving abilities for shorter RNAs. In our experiments, we found the maximum capacity of a 1 mm thick APB gel with a 1 cm wide well was 2 μg of total RNA after which the retention of unmodified RNAs was skewed. When dealing with minute tissue samples, however, total RNA is typically limiting, such as where a single follicle cell enriched sample might typically yield 5 μg of total RNA. When electrophoresing an RNA sample on the APB gel, we recommend resolving no more than 2 μg of RNA for a 1 cm wide, 0.8 mm thick gel to avoid overloading the capacity of the APB. Since modified RNAs run very differently compared to unmodified RNAs of the same size, we recommend that a standard synthetic RNA and a 2′-*O*-methylated RNA oligo of similar size to piRNAs be radiolabeled and run alongside your organismal RNA sample to guide where to cut out the gel region of interest from which to elute the RNA in 0.5 M NaCl for downstream small RNA library construction.
3. Although sequencing chemistry for the Illumina platform continues to improve in providing longer read lengths to over 100 nt, the fidelity and quality of reading the first couple of bases extending from a sequencing or indexing primer are still considerably higher than readings towards the end of the molecule. Thus, we have chosen to place our linker barcode sequence at the first set of bases to be read by the small RNA

sequencing primer #1, in contrast to other methodologies that place the barcode sequence in the 3′ Linker [12, 13]. On the Illumina Genome Analyzer, the platform uses approximately 8 bases of read information to fine tune the assign and distinguish the "molecule colonies" on the flow cell lane, and our original 4 base barcode was suitable for multiplexing. However, the Hi-Seq platform only utilizes the initial 4–6 bases, and thus finds the original 4 base barcode too low in complexity to distinguish colonies, thus drastically reducing the theoretical numbers of quality-passing reads by threefold (Fig. 1f).

Acknowledgments

J.A.M and Y.S. contributed equally to this work. We are grateful to Kuniaki Saito and Haruhiko and Mikiko Siomi for antibody reagents to *Drosophila* Piwi proteins and to Dorothea Godt for the antibody to TJ. We thank Michael Rosbash and his laboratory members for access to the Illumina sequencer. This work was supported by the National Institutes of Health (Core Facilities Grant P30 NS045713 to the Brandeis Biology Department and R00HD057298 to N.C.L.). N.C.L. is a Searle Scholar.

References

1. Lau NC, Lim LP, Weinstein EG, Bartel DP (2001) An abundant class of tiny RNAs with probable regulatory roles in *Caenorhabditis elegans*. Science 294:858–862
2. Lee RC, Ambros V (2001) An extensive class of small RNAs in *Caenorhabditis elegans*. Science 294:862–864
3. Lagos-Quintana M, Rauhut R, Lendeckel W, Tuschl T (2001) Identification of novel genes coding for small expressed RNAs. Science 294:853–858
4. Lau NC, Robine N, Martin R, Chung WJ, Niki Y, Berezikov E, Lai EC (2009) Abundant primary piRNAs, endo-siRNAs, and microRNAs in a Drosophila ovary cell line. Genome Res 19:1776–1785
5. Lau NC, Ohsumi T, Borowsky M, Kingston RE, Blower MD (2009) Systematic and single cell analysis of Xenopus Piwi-interacting RNAs and Xiwi. EMBO J 28:2945–2958
6. Lau NC (2008) Analysis of small endogenous RNAs. Curr Protoc Mol Biol. Chapter 26, Unit26 7
7. Saito K, Sakaguchi Y, Suzuki T, Siomi H, Siomi MC (2007) Pimet, the Drosophila homolog of HEN1, mediates 2′-O-methylation of Piwi-interacting RNAs at their 3′ ends. Genes Dev 21:1603–1608
8. Horwich MD, Li C, Matranga C, Vagin V, Farley G, Wang P, Zamore PD (2007) The Drosophila RNA methyltransferase, DmHen1, modifies germline piRNAs and single-stranded siRNAs in RISC. Curr Biol 17:1265–1272
9. Igloi GL, Kossel H (1987) Use of boronate-containing gels for electrophoretic analysis of both ends of RNA molecules. Methods Enzymol 155:433–448
10. Babiarz JE, Ruby JG, Wang Y, Bartel DP, Blelloch R (2008) Mouse ES cells express endogenous shRNAs, siRNAs, and other Microprocessor-independent, Dicer-dependent small RNAs. Genes Dev 22:2773–2785
11. Chiang HR, Schoenfeld LW, Ruby JG, Auyeung VC, Spies N, Baek D, Johnston WK, Russ C, Luo S, Babiarz JE, Blelloch R, Schroth GP, Nusbaum C, Bartel DP (2010) Mammalian microRNAs: experimental evaluation of novel and previously annotated genes. Genes Dev 24:992–1009
12. Hafner M, Landgraf P, Ludwig J, Rice A, Ojo T, Lin C, Holoch D, Lim C, Tuschl T (2008) Identification of microRNAs and other small regulatory RNAs using cDNA library sequencing. Methods 44:3–12
13. Hafner M, Renwick N, Farazi TA, Mihailovic A, Pena JT, Tuschl T (2012) Barcoded cDNA library

preparation for small RNA profiling by next-generation sequencing. Methods 58:164–170
14. Ruby JG, Stark A, Johnston WK, Kellis M, Bartel DP, Lai EC (2007) Evolution, biogenesis, expression, and target predictions of a substantially expanded set of Drosophila microRNAs. Genome Res 17:1850–1864
15. Ro S, Yan W (2010) Small RNA cloning. Methods Mol Biol 629:273–285
16. Thomas MF, Ansel KM (2010) Construction of small RNA cDNA libraries for deep sequencing. Methods Mol Biol 667:93–111
17. Havecker ER (2011) Detection of small RNAs and microRNAs using deep sequencing technology. Methods Mol Biol 732:55–68
18. Donovan WP, Zhang Y, Howell MD (2011) Large-scale sequencing of plant small RNAs. Methods Mol Biol 744:159–173
19. Claycomb JM, Benasutti M, Bosco G, Fenger DD, Orr-Weaver TL (2004) Gene amplification as a developmental strategy: isolation of two developmental amplicons in Drosophila. Dev Cell 6:145–155
20. Bryant Z, Subrahmanyan L, Tworoger M, LaTray L, Liu CR, Li MJ, van den Engh G, Ruohola-Baker H (1999) Characterization of differentially expressed genes in purified Drosophila follicle cells: toward a general strategy for cell type-specific developmental analysis. Proc Natl Acad Sci U S A 96:5559–5564

Chapter 11

Analysis of sDMA Modifications of PIWI Proteins

Shozo Honda, Yoriko Kirino, and Yohei Kirino

Abstract

Arginine methylation is an important posttranslational protein modification that modulates protein function for a wide range of biological processes. PIWI proteins, a subclade of the Argonaute family proteins, contain evolutionarily conserved symmetrical dimethylarginines (sDMAs). It has become increasingly apparent that the sDMAs of PIWI proteins serve as binding elements for TUDOR domain-containing proteins and that sDMA-dependent protein interactions play crucial roles in the biogenesis and function of PIWI-interacting RNAs (piRNAs). We describe a method for detecting PIWI sDMAs and purifying PIWI/piRNA complexes using anti-sDMA antibodies.

Key words PIWI, piRNA, Arginine methylation, Symmetrical dimethylarginine (sDMA), Y12, SYM10, SYM11

1 Introduction

Arginine methylation is an important posttranslational protein modification that plays crucial roles in numerous biological processes, such as structural remodeling of chromatin, signal transduction, mRNA splicing, and DNA repair [1–5]. Arginine methylation is mediated by two types of protein methyltransferases (PRMTs): type I enzymes (e.g., PRMT1) catalyze asymmetrical dimethylarginine (aDMA) in which two methyl groups are placed on one of the terminal nitrogen atoms of the guanidino group, whereas type II enzymes (e.g., PRMT5) catalyze symmetrical dimethylarginine (sDMA) (Fig. 1a) in which one methyl group is placed on each of the terminal nitrogens [1–5]. sDMA modifications occur in "sDMA motifs" comprising arginines flanked by glycines (GRG) or alanines (GRA or ARG), which are often found as repeats. sDMAs are known to specifically bind to TUDOR domains of proteins and regulate protein–protein interactions [1–5]. For example, in mammals, sDMAs of Sm proteins, components of small nuclear ribonucleoproteins (snRNPs), promote their binding to

Mikiko C. Siomi (ed.), *PIWI-Interacting RNAs: Methods and Protocols*, Methods in Molecular Biology, vol. 1093, DOI 10.1007/978-1-62703-694-8_11,

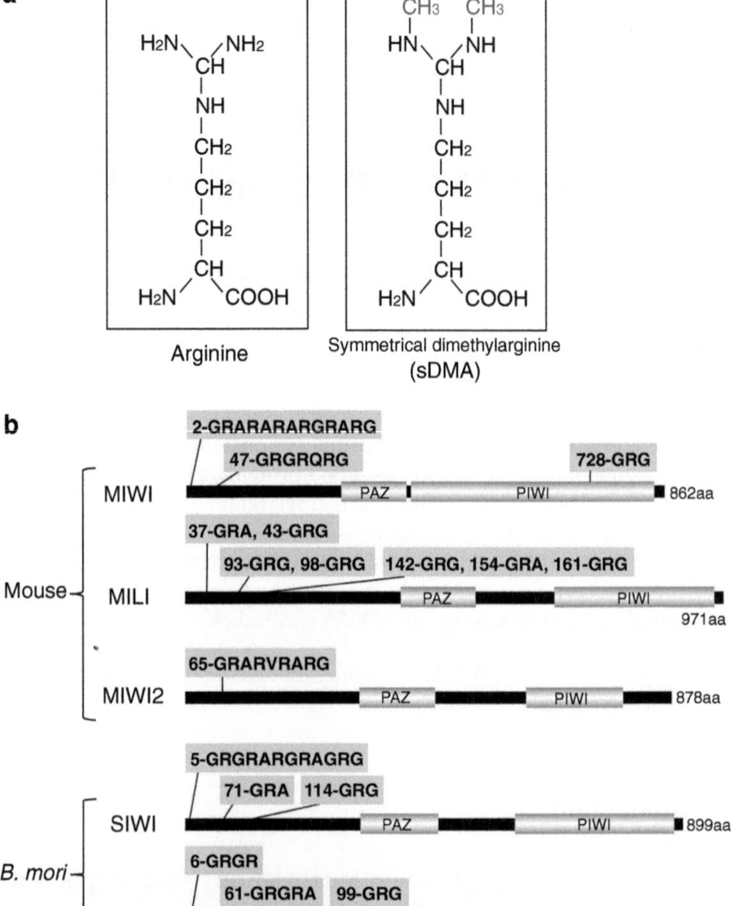

Fig. 1 sDMA motifs in PIWI proteins. (**a**) Chemical structure of arginine and sDMA. (**b**) The structures of PIWI proteins (MIWI, MILI, and MIWI2 for mouse; SIWI and BmAGO3 for *Bombyx mori*) are shown with sDMA motifs comprising GRG or GRA/ARG sequences. PAZ and PIWI are the two major protein motifs that exist in the Argonaute family proteins

the TUDOR domain of survival of motor neuron (SMN) protein, which facilitates snRNP assembly [6, 7].

Arginine methylation and its mediated protein–protein interactions have attracted increased attention as key molecular factors in small regulatory RNA pathway in the germline. PIWI proteins, a subclade of the Argonaute family proteins, are predominantly expressed in the germline and bind to 25–31 nucleotide (nt) PIWI-interacting RNAs (piRNAs) to form PIWI-ribonucleoproteins (piRNPs) [8–10]. Mice express three PIWI proteins: MIWI, MILI, and MIWI2 [11–13]; and *Bombyx mori* (silkworm) expresses two PIWI proteins: SIWI and BmAGO3 [14]. piRNPs play critical roles in germline development by regulating transposons and other targets to maintain genome integrity. PIWI proteins contain evolutionarily conserved sDMAs that are synthesized by PRMT5 [15]. Putative sDMA motifs are typically clustered at the N-terminus of PIWI proteins (Fig. 1b); in mouse and *Drosophila*, sDMA positions were further determined by mass spectrometry [16–19]. Several members of the TUDOR domain-containing protein family, such as Spindle-E, Tudor, Krimper, and Tejas in *Drosophila* [19–23] and TDRD1-9 in mice [16–18, 24–26], have recently been studied for their PIWI interactions and functional involvements in piRNA biogenesis and function [27, 28].

Here, we describe a method for detecting sDMA modifications of PIWI proteins. There are three highly specific anti-sDMA antibodies: Y12, SYM10, and SYM11. Y12 is a monoclonal antibody that was generated from a hybridoma of lupus erythmatosus-like syndrome mice developing autoantibodies against Sm proteins [29, 30]. The epitope that Y12 recognizes on Sm proteins comprises sDMAs of the proteins [31]. SYM10 and SYM11 are polyclonal antibodies derived from rabbit serum immunized with peptides containing sDMAs, K(sDMA)G(sDMA)G(sDMA)G(sDMA)G and KAAILKAQVAA(sDMA)G(sDMA)G(sDMA)GMG(sDMA)G, respectively [32, 33]. We describe a utilization of these antibodies to detect sDMAs of MIWI and MILI, which were purified from mouse testicles, and SIWI and BmAGO3, which were transiently expressed and purified from BmN4, a *Bombyx mori* ovary-derived cultured cell line [34].

In addition, we describe a method to purify piRNP using immunoprecipitation with the Y12 antibody. For piRNA purification and identification, piRNPs are typically purified by anti-PIWI immunoprecipitation. However, we previously reported successful piRNP purification using Y12 immunoprecipitation for mouse testicles and Xenopus oocytes [15]; we here demonstrate that it can also be used for BmN4 cells. Our results suggest that the Y12 antibody can be widely used to purify piRNPs for identifying piRNA sequences in various organisms for which antibodies against PIWI proteins have not yet been generated.

2 Materials

1. Recombinant protein G agarose beads (Invitrogen).
2. Anti-Flag® M2-agarose from mouse (Sigma).
3. Y12 antibody (mouse monoclonal; a gift from G. Dreyfuss, University of Pennsylvania) (*see* **Note 1**).
4. SYM10 antibody (rabbit polyclonal) (Millipore).
5. SYM11 antibody (rabbit polyclonal) (Millipore).

6. Anti-MIWI antibody (rabbit polyclonal) [23] (*see* **Note 1**).
7. Anti-MILI antibody (mouse monoclonal; clone 17.8) [15] (*see* **Note 1**).
8. Nonimmune mouse and rabbit serum.
9. Mouse testicle (Pel-Freez Biochemicals).
10. Lysis buffer: 20 mM Tris–HCl (pH 7.4); 200 mM NaCl; 2.5 mM $MgCl_2$; 0.5 % NP-40; 0.1 % Triton X-100; one tablet of Complete protease inhibitor EDTA-free (Roche) per 50 mL of lysis buffer.
11. 7 mL Dounce tissue grinder (Wheaton).
12. Bioruptor sonication system (Diagenode).
13. BmN4 cell line (a gift from S. Katsuma, The University of Tokyo).
14. Expression plasmids for Flag-SIWI and Flag-BmAGO3: The N-terminal Flag/His-tagged SIWI or BmAGO3 were cloned into a pIZ/V5-His vector (a gift from S. Katsuma, The University of Tokyo) [34].
15. Insect X-press medium (LONZA).
16. Sf-900™ III SFM (1×), liquid (Invitrogen).
17. ESCORT transfection reagent (Sigma).
18. NuPAGE LDS sample buffer (Invitrogen).
19. β-Mercaptoethanol.
20. NuPAGE 4–12 % Bis-Tris gel (Invitrogen).
21. NuPAGE MOPS SDS running buffer (Invitrogen).
22. SilverQuest staining kit (Invitrogen).
23. Nitrocellulose/filter paper; 0.45 μm pore size (Invitrogen).
24. Transfer buffer: 62.5 mM Tris, 18 mM Glycine, 20 % Methanol.
25. TE70 ECL semi-dry transfer unit (GE Healthcare).
26. PBS.
27. PBST: PBS containing 0.1 % Tween 20.
28. Blocking solution: 5 % nonfat dry milk in PBST.
29. ECL anti-rabbit IgG, horseradish peroxidase linked F(ab′) 2 fragment from donkey (GE Healthcare).
30. ECL anti-mouse IgG, horseradish peroxidase linked F(ab′) 2 fragment from sheep (GE Healthcare).
31. ECL plus western blotting detection system (GE Healthcare).
32. ChemiDoc™ XRS+ system (Bio-Rad).
33. Trizol (Invitrogen).
34. Glycogen (Ambion).
35. 3 M NaOAc (pH 5.5) (Ambion).

36. Isopropanol.
37. Centrifugal evaporator (myVac).
38. Alkaline phosphatase, calf intestinal (CIP) (NEB).
39. T4 Polynucleotide Kinase (T4 PNK) (NEB).
40. [γ-32P] ATP (American Radiolabeled Chemicals).
41. 15 % PAGE solution with 7 M Urea: 420.42 g Urea, 376 mL 40 % acrylamide and bis-acrylamide solution (19:1) (Bio-Rad); 100 mL Ultrapure™ 10× TBE buffer (Invitrogen) and MilliQ water to prepare 1 L. After filtration, store at 4 °C in dark.
42. Ammonium persulfate.
43. Ultrapure™ TEMED (Invitrogen).
44. SE-400 electrophoresis system (Hoefer).
45. 2× Loading buffer for Urea PAGE: 5.4 g Urea, 6 mg Bromophenol blue, 6 mg Xylene cyanol, and MilliQ water for 10 mL.
46. Phosphor autoradiography plate (Kodak).
47. Molecular Imager PharosFX System (Bio-Rad).

3 Methods

3.1 Purification of MIWI and MILI from Mouse Testicles by Immunoprecipitation

3.1.1 Preparation of Antibody-Bound Agarose Beads

1. Wash protein G agarose beads (10 μL bed volume) three times with 1 mL of lysis buffer.
2. Add either anti-MIWI (10 μL), anti-MILI (2.5 μL), nonimmune mouse serum (NMS, negative control), or nonimmune rabbit serum (NRS, negative control) to the beads in 700 μL of lysis buffer.
3. Rotate for 1 h at room temperature (RT).
4. Discard the buffer containing antibody and wash five times with 1 mL of lysis buffer.

3.1.2 Preparation of Mouse Testicle Lysate

1. Use one mouse testicle per immunoprecipitation (500 μL of lysis buffer). Homogenize testicles in lysis buffer with a Dounce tissue grinder in a cold room (4 °C).
2. Sonicate the homogenate using the Biorupter sonication system according to manufacturer's instructions (on: 5 s, off: 7 s, cycles: 14, strength level: medium).
3. Centrifuge the lysate at 20,000 × *g* for 10 min at 4 °C. Collect the supernatant.

3.1.3 Immunoprecipitation

1. Add the testicle lysate to the prepared beads with antibodies and rotate for 1 h at 4 °C.
2. Wash the beads five times with 1 mL of lysis buffer.

3.1.4 Visualization of Purified Proteins on SDS-PAGE by Silver Staining

1. Add 30 μL of NuPAGE LDS sample buffer containing 10 % β-mercaptoethanol to the beads and incubate at 70 °C for 15 min.
2. Run 10 μL of immunoprecipitate samples on NuPAGE 4–12 % Bis-Tris gel.
3. Stain with SilverQuest staining kit according to the manufacturer's instructions (Fig. 2a).

3.2 Purification of SIWI and BmAGO3 from BmN4 Cells by Immunoprecipitation

3.2.1 Transient Expression of Flag-SIWI and Flag-BmAGO3 in BmN4 Cells

1. Spread and culture 5×10^6 BmN4 cells on a 10 cm dish in Insect X-press medium at 27 °C for 24 h, then change the medium with 5 mL of Sf-900 medium.
2. Gently mix 13 μg of expression plasmid with 40 μL of ESCORT reagent and 650 μL of Sf-900 medium by pipetting, and then incubate at RT for 15 min.
3. Add this mixture to cells, incubate at 27 °C for 6 h, and add 5 mL of Sf-900 medium.
4. After 3 days, collect cells with a scraper and wash the cells twice with 1 mL of PBS.

3.2.2 Preparation of BmN4 Cell Lysate

1. Resuspend cells in lysis buffer (5×10^6 cells in 200 μL of lysis buffer per immunoprecipitation).
2. Sonicate the suspension and collect the supernatant as described in Subheading 3.1.2.

3.2.3 Anti-flag Immunoprecipitation

1. Wash anti-Flag agarose beads (10 μL bed volume) three times with 1 mL of lysis buffer.
2. Add the 200 μL of lysate to the beads and adjust total volume to 500 μL with lysis buffer so that the beads are not stacked within the tube during rotation.
3. Rotate for 1 h at 4 °C, and then prepare samples for SDS-PAGE as described in Subheading 3.1.4.

3.3 Detection of PIWI sDMAs by Western Blots Using Anti-sDMA Antibodies

3.3.1 SDS-PAGE and Transfer

1. Run 4 μL of immunoprecipitate samples on NuPAGE 4–12 % Bis-Tris gel.
2. Immerse the gel, nitrocellulose membrane and filter paper in transfer buffer and perform transfer procedure with TE70 ECL semi-dry transfer unit according to the manufacturer's instructions (run at 90 mA for 45 min).

3.3.2 Immunoblotting and Detection

1. Rinse the membrane with MilliQ water, and then incubate the membrane with blocking buffer at 70 rpm for 1 h at RT.
2. Discard the blocking buffer and rinse the membrane twice with PBST.

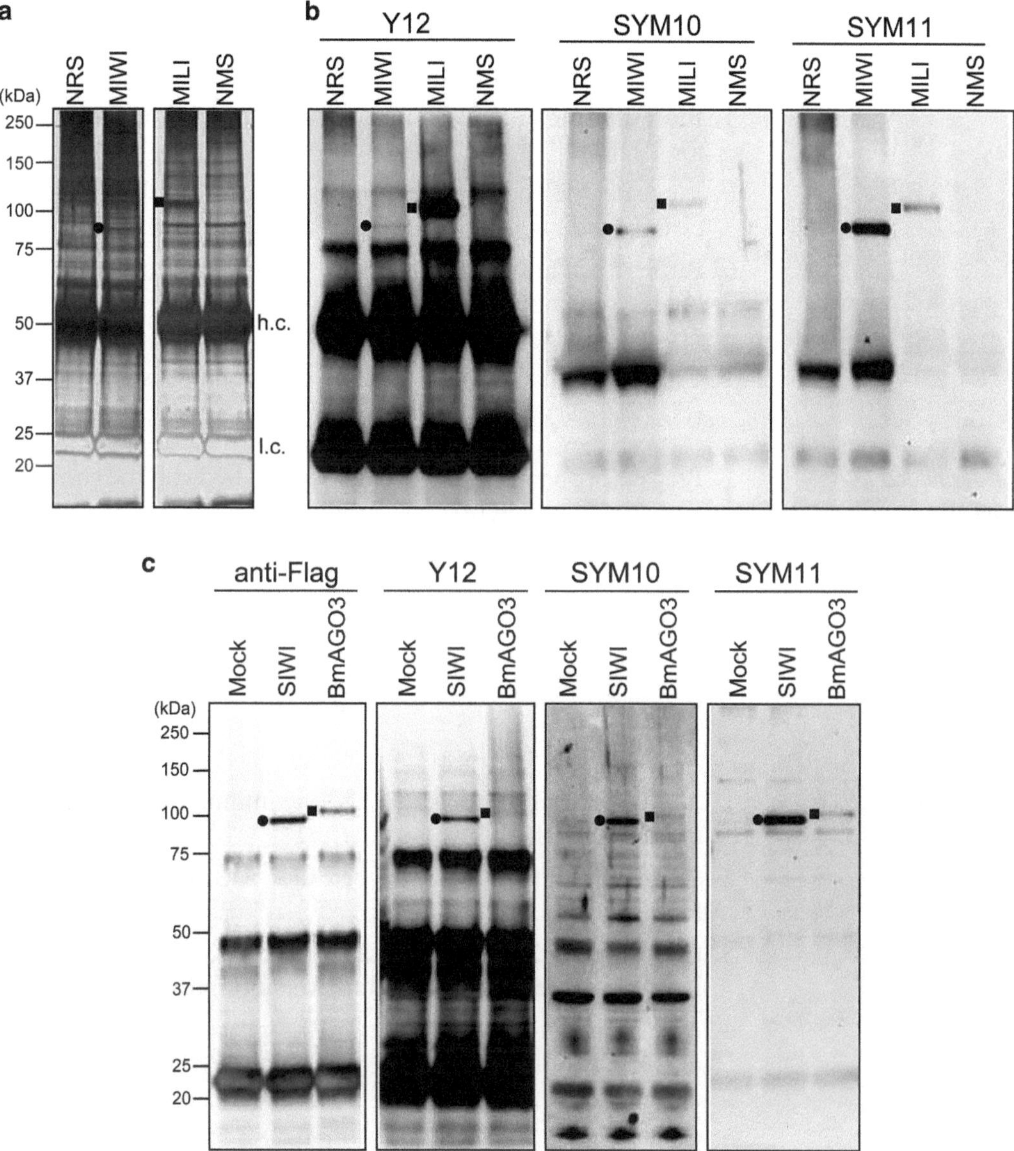

Fig. 2 Detection of PIWI sDMAs by western blot with anti-sDMA antibodies. (**a**) MIWI and MILI immunoprecipitates from mouse testicles were visualized by silver staining (*NMS* nonimmune mouse serum, *NRS* nonimmune rabbit serum). The *filled circle* and *square* indicate the bands for purified MIWI and MILI, respectively. *h.c.* and *l.c.* indicate antibody heavy and light chains, respectively. (**b**) Immunoprecipitates from mouse testicles were probed on western blots with the indicated antibodies. The Y12, SYM10, and SYM11 antibodies recognized both MIWI (*filled circle*) and MILI (*filled square*), which indicated that MIWI and MILI contain sDMAs. (**c**) Flag-SIWI or Flag-BmAGO3 expressed in BmN4 cells was immunopurified with an anti-Flag antibody (Mock: the sample from BmN4 cells with no transient protein expression) and probed on western blots with the indicated antibodies. All these antibodies recognized SIWI (*filled circle*) and BmAGO3 (*filled square*), indicating the presence of sDMAs in SIWI and BmAGO3

3. Incubate the membrane with a primary antibody solution (anti-Flag 1:1,000; Y12 1:500; SYM10 1:1,000; and SYM11 1:1,000; each diluted in blocking buffer) at 4 °C overnight.
4. Rinse the membrane briefly and wash three times with PBST for 10 min.
5. Incubate the membrane with a secondary antibody solution (1:5,000; diluted in PBST) at RT for 1 h.
6. Rinse the membrane with PBST, wash three times with PBST for 10 min, and immerse in PBS.
7. Detect bands using ECL-Plus detection solution and ChemiDoc according to the manufacturer's instructions (Fig. 2b, c) (*see* **Note 2**).

3.4 Isolation of piRNPs from BmN4 Cells by Immunoprecipitation with Y12 Antibody

3.4.1 Immuno-precipitation with Y12 Antibody

1. Prepare protein G agarose beads (10 μL bed volume) bound to Y12 (5 μL) as described in Subheading 3.1.1.
2. Prepare BmN4 cell lysate (1×10^7 cells per immunoprecipitation) as described in Subheading 3.2.2.
3. Add the lysate to the beads and rotate at 4 °C for 2 h.
4. Wash the beads five times with 1 mL of lysis buffer.

3.4.2 Isolation of piRNAs

1. Add 500 μL of Trizol reagent to the immunoprecipitate beads and suspend the beads by pipetting.
2. Add 100 μL of chloroform, vortex for 15 s, and let stand at RT for 2 min.
3. Centrifuge at 20,000 × *g* for 30 min at 4 °C. Collect the upper aqueous phase (carefully exclude the interphase).
4. Add 2 μL of glycogen (5 mg/mL), vortex briefly, add 350 μL of isopropanol, and vortex again. Cool the tube at −20 °C for 20 min.
5. Centrifuge at 20,000 × *g* for 30 min at 4 °C. Carefully remove the supernatant and dry the pellet for 1 min with a centrifugal evaporator.
6. Dissolve the pellet in 14 μL of MilliQ water with thorough pipetting.
7. Store the piRNA solution at −80 °C.

3.4.3 5′-End Radiolabeling of Isolated piRNAs

1. To remove the 5′-end phosphate of piRNAs, incubate 7 μL of the isolated piRNA solution with 0.5 μL of CIP, 2 μL of 10× buffer, and 10.5 μL MilliQ water (total 20 μL) at 37 °C for 30 min.
2. Adjust the total volume to 100 μL with MilliQ water, add 100 μL of phenol and thoroughly vortex.

3. Centrifuge at 20,000 × *g* for 5 min at RT and then carefully collect the upper phase.
4. Add 2 μL of glycogen (5 mg/mL), 10 μL of 3 M NaOAc and briefly vortex. Then, add 275 μL of chilled 100 % ethanol, vortex well and cool the tube at −80 °C for 30 min.
5. Centrifuge at 20,000 × *g* for 30 min at 4 °C. Carefully remove the supernatant and dry the pellet for 1 min with a centrifugal evaporator.
6. Dissolve the pellet with 7.5 μL of MilliQ water with thorough pipetting.
7. For 5′-end labeling of piRNAs, incubate 7.5 μL of dephosphorylated piRNA solution with 1 μL of [γ-32P]ATP, 0.5 μL of T4 PNK, and 1 μL of 10× buffer (total 10 μL) for 1 h at 37 °C.

3.4.4 Separation and Detection of Labeled piRNAs

1. Set up the gel apparatus (SE-400 system) using 18 × 24 cm glass plate and 0.75 mm thick combs.
2. Add 150 μL of 10 % ammonium persulfate and 15 μL of TEMED to 30 mL of 15 % 7 M Urea PAGE solution. Mix gently and immediately pour into the apparatus.
3. Run the gel at 300 V for 30 min and wash the wells with a syringe to remove accumulated urea in the wells.
4. Add an equal volume of 2× loading buffer to the labeled piRNA solution.
5. Run 10 μL of the samples at 700 V until the dye (BPB) front reaches the bottom of the gel.
6. Disassemble glass plates and peel out one side of glass plate with a spatula. Cover the gel with a wrap and expose the gel to a phosphor autoradiography plate in a cassette for 1 h to overnight at −80 °C.
7. Scan the plate using a PharosFX phosphor imager (Fig. 3) (*see* **Notes 2–4**).

4 Notes

1. The Y12, anti-MIWI, and anti-MILI antibodies are commercially available from various distributors.
2. The affinities of the respective anti-sDMAs for PIWI proteins have certain differences. SYM10 and SYM11 recognize MIWI and SIWI well. This is because both of these PIWI proteins contain a long GRG/ARG/GRA repeat (Fig. 1b) that is similar to the sequences of the peptides used for producing SYM10 and SYM11 [32, 33]. In contrast, the Y12 antibody attaches to

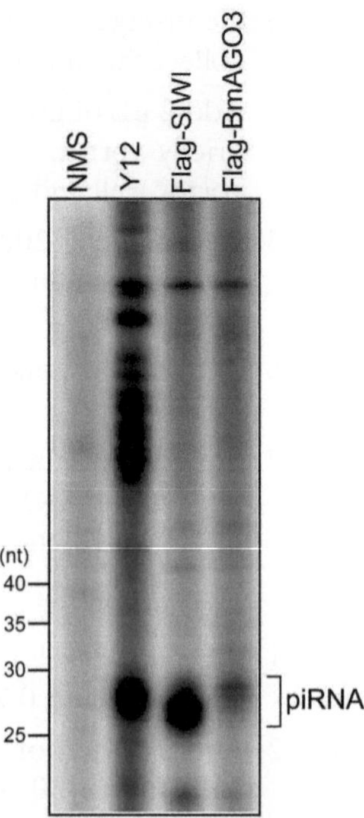

Fig. 3 piRNP purification by immunoprecipitation with the Y12 antibody. Y12 immunoprecipitate from BmN4 cells, and anti-Flag immunoprecipitates from BmN4 cells expressing Flag-SIWI or Flag-BmAGO3, were subjected to RNA extraction, 5′-end labeling of the extracted RNA, and denaturing PAGE. piRNAs were clearly observed in Y12 immunoprecipitate as well as PIWIs, which indicated the successful purification of piRNPs with the Y12 antibody

MILI much more strongly than to MIWI on western blot (Fig. 2b) and in immunoprecipitation [15].

3. It has been reported that immunoprecipitation with SYM10 or SYM11 could be used to pull down various sDMA-containing proteins, including Sm proteins [32, 33, 35]. Based on these reports, we attempted to perform SYM10 and SYM11 immunoprecipitation using lysates from mouse testicles or Drosophila ovaries to purify piRNPs. However, despite using different lysis buffers containing different salt concentrations (100–200 mM), we were unable to detect PIWI proteins and piRNAs in these immunoprecipitates. It was observed that, among the anti-sDMA antibodies, the Y12 antibody was particularly useful for purifying piRNPs.
4. Y12 purifies snRNPs containing snRNAs [29] as well as piRNPs containing piRNAs. Therefore, the long RNAs observed in Y12 immunoprecipitate (Fig. 3) include snRNAs.

Acknowledgments

We are grateful to G. Dreyfuss for the Y12 antibody, to S. Katsuma for BmN4 and PIWI expression constructs, and to Z. Mourelatos for support and discussion. This work was supported by the Cedars-Sinai Medical Center Research Fund, Martz Translational Breast Cancer Research Fund, and a Grant for Basic Science Research Project from The Sumitomo Foundation (Y.K.).

References

1. Bedford MT, Richard S (2005) Arginine methylation an emerging regulator of protein function. Mol Cell 18:263–272
2. Krause CD, Yang ZH, Kim YS, Lee JH, Cook JR, Pestka S (2007) Protein arginine methyltransferases: evolution and assessment of their pharmacological and therapeutic potential. Pharmacol Ther 113:50–87
3. Wolf SS (2009) The protein arginine methyltransferase family: an update about function, new perspectives and the physiological role in humans. Cell Mol Life Sci 66:2109–2121
4. Bedford MT, Clarke SG (2009) Protein arginine methylation in mammals: who, what, and why. Mol Cell 33:1–13
5. Blackwell E, Ceman S (2012) Arginine methylation of RNA-binding proteins regulates cell function and differentiation. Mol Reprod Dev 79:163–175
6. Yong J, Wan L, Dreyfuss G (2004) Why do cells need an assembly machine for RNA-protein complexes? Trends Cell Biol 14:226–232
7. Meister G, Eggert C, Fischer U (2002) SMN-mediated assembly of RNPs: a complex story. Trends Cell Biol 12:472–478
8. Ghildiyal M, Zamore PD (2009) Small silencing RNAs: an expanding universe. Nat Rev Genet 10:94–108
9. Kim VN, Han J, Siomi MC (2009) Biogenesis of small RNAs in animals. Nat Rev Mol Cell Biol 10:126–139
10. Farazi TA, Juranek SA, Tuschl T (2008) The growing catalog of small RNAs and their association with distinct Argonaute/Piwi family members. Development 135:1201–1214
11. Kuramochi-Miyagawa S, Kimura T, Yomogida K, Kuroiwa A, Tadokoro Y, Fujita Y, Sato M, Matsuda Y, Nakano T (2001) Two mouse piwi-related genes: miwi and mili. Mech Dev 108:121–133
12. Deng W, Lin H (2002) miwi, a murine homolog of piwi, encodes a cytoplasmic protein essential for spermatogenesis. Dev Cell 2:819–830
13. Carmell MA, Girard A, van de Kant HJ, Bourc'his D, Bestor TH, de Rooij DG, Hannon GJ (2007) MIWI2 is essential for spermatogenesis and repression of transposons in the mouse male germline. Dev Cell 12:503–514
14. Kawaoka S, Minami K, Katsuma S, Mita K, Shimada T (2008) Developmentally synchronized expression of two Bombyx mori Piwi subfamily genes, SIWI and BmAGO3 in germ-line cells. Biochem Biophys Res Commun 367:755–760
15. Kirino Y, Kim N, de Planell-Saguer M, Khandros E, Chiorean S, Klein PS, Rigoutsos I, Jongens TA, Mourelatos Z (2009) Arginine methylation of Piwi proteins catalysed by dPRMT5 is required for Ago3 and Aub stability. Nat Cell Biol 11:652–658
16. Reuter M, Chuma S, Tanaka T, Franz T, Stark A, Pillai RS (2009) Loss of the Mili-interacting Tudor domain-containing protein-1 activates transposons and alters the Mili-associated small RNA profile. Nat Struct Mol Biol 16:639–646
17. Vagin VV, Wohlschlegel J, Qu J, Jonsson Z, Huang X, Chuma S, Girard A, Sachidanandam R, Hannon GJ, Aravin AA (2009) Proteomic analysis of murine Piwi proteins reveals a role for arginine methylation in specifying interaction with Tudor family members. Genes Dev 23:1749–1762
18. Chen C, Jin J, James DA, Adams-Cioaba MA, Park JG, Guo Y, Tenaglia E, Xu C, Gish G, Min J, Pawson T (2009) Mouse Piwi interactome identifies binding mechanism of Tdrkh Tudor domain to arginine methylated Miwi. Proc Natl Acad Sci U S A 106:20336–20341
19. Nishida KM, Okada TN, Kawamura T, Mituyama T, Kawamura Y, Inagaki S, Huang H, Chen D, Kodama T, Siomi H, Siomi MC (2009) Functional involvement of Tudor and dPRMT5 in the piRNA processing pathway in Drosophila germlines. EMBO J 28:3820–3831
20. Vagin VV, Sigova A, Li C, Seitz H, Gvozdev V, Zamore PD (2006) A distinct small RNA pathway silences selfish genetic elements in the germline. Science 313:320–324

21. Lim AK, Kai T (2007) Unique germ-line organelle, nuage, functions to repress selfish genetic elements in Drosophila melanogaster. Proc Natl Acad Sci U S A 104:6714–6719
22. Patil VS, Kai T (2010) Repression of retroelements in Drosophila germline via piRNA pathway by the Tudor domain protein Tejas. Curr Biol 20:724–730
23. Kirino Y, Vourekas A, Sayed N, de Lima Alves F, Thomson T, Lasko P, Rappsilber J, Jongens TA, Mourelatos Z (2010) Arginine methylation of Aubergine mediates Tudor binding and germ plasm localization. RNA 16:70–78
24. Wang J, Saxe JP, Tanaka T, Chuma S, Lin H (2009) Mili interacts with tudor domain-containing protein 1 in regulating spermatogenesis. Curr Biol 19:640–644
25. Kojima K, Kuramochi-Miyagawa S, Chuma S, Tanaka T, Nakatsuji N, Kimura T, Nakano T (2009) Associations between PIWI proteins and TDRD1/MTR-1 are critical for integrated subcellular localization in murine male germ cells. Genes Cells 14:1155–1165
26. Shoji M, Tanaka T, Hosokawa M, Reuter M, Stark A, Kato Y, Kondoh G, Okawa K, Chujo T, Suzuki T, Hata K, Martin SL, Noce T, Kuramochi-Miyagawa S, Nakano T, Sasaki H, Pillai RS, Nakatsuji N, Chuma S (2009) The TDRD9-MIWI2 complex is essential for piRNA-mediated retrotransposon silencing in the mouse male germline. Dev Cell 17:775–787
27. Siomi MC, Mannen T, Siomi H (2010) How does the royal family of Tudor rule the PIWI-interacting RNA pathway? Genes Dev 24:636–646
28. Chen C, Nott TJ, Jin J, Pawson T (2011) Deciphering arginine methylation: Tudor tells the tale. Nat Rev Mol Cell Biol 12:629–642
29. Lerner EA, Lerner MR, Janeway CA Jr, Steitz JA (1981) Monoclonal antibodies to nucleic acid-containing cellular constituents: probes for molecular biology and autoimmune disease. Proc Natl Acad Sci U S A 78:2737–2741
30. Pisetsky DS, Lerner EA (1982) Idiotypic analysis of a monoclonal anti-Sm antibody. J Immunol 129:1489–1492
31. Brahms H, Raymackers J, Union A, de Keyser F, Meheus L, Luhrmann R (2000) The C-terminal RG dipeptide repeats of the spliceosomal Sm proteins D1 and D3 contain symmetrical dimethylarginines, which form a major B-cell epitope for anti-Sm autoantibodies. J Biol Chem 275:17122–17129
32. Boisvert FM, Cote J, Boulanger MC, Cleroux P, Bachand F, Autexier C, Richard S (2002) Symmetrical dimethylarginine methylation is required for the localization of SMN in Cajal bodies and pre-mRNA splicing. J Cell Biol 159:957–969
33. Boisvert FM, Cote J, Boulanger MC, Richard S (2003) A proteomic analysis of arginine-methylated protein complexes. Mol Cell Proteomics 2:1319–1330
34. Kawaoka S, Hayashi N, Suzuki Y, Abe H, Sugano S, Tomari Y, Shimada T, Katsuma S (2009) The Bombyx ovary-derived cell line endogenously expresses PIWI/PIWI-interacting RNA complexes. RNA 15:1258–1264
35. Zhang Z, Zhang S, Zhang Y, Wang X, Li D, Li Q, Yue M, Zhang YE, Xu Y, Xue Y, Chong K, Bao S (2011) Arabidopsis floral initiator SKB1 confers high salt tolerance by regulating transcription and pre-mRNA splicing through altering histone H4R3 and small nuclear ribonucleoprotein LSM4 methylation. Plant Cell 23:396–411

Chapter 12

Analyses of piRNA-Mediated Transcriptional Transposon Silencing in *Drosophila*: Nuclear Run-On Assay on Ovaries

Sergey Shpiz and Alla Kalmykova

Abstract

In the *Drosophila* germline, retrotransposons are silenced by the PIWI-interacting RNA (piRNA) pathway. piRNA pathway mutations lead to overexpression and mobilization of retrotransposons in the germline. In different organisms, small RNAs were shown to be implicated in the posttranscriptional degradation of mRNA and/or transcriptional repression of the homologous locus. In *Drosophila*, the mechanism of piRNA-mediated silencing is still far from being understood. Transcriptional silencing implies a piRNA-mediated formation of repressive chromatin which diminishes the transcriptional capacity of the target locus. Nuclear Run-On (NRO) assay allows a direct estimation of the density of transcribing polymerases at specific genomic regions. Here we describe the NRO protocol on *Drosophila* ovarian tissues which can be useful for investigation of the transcriptional silencing in the female germline.

Key words piRNA, Gene silencing, Transcription, Transposon, Chromatin, Nuclear run-on, *Drosophila*

1 Introduction

piRNAs associated with Argonaute proteins from the PIWI subfamily protect the germline against the transposable elements activity [1]. In different organisms, small RNAs were shown to be implicated in posttranscriptional silencing and/or heterochromatin formation at the target locus. Studies in fission yeast, plants, and ciliates have revealed a role for small RNAs in methylation of lysine 9 of histone 3 (H3K9) with a subsequent binding of heterochromatic protein 1 homologues [2–4]. In the *Drosophila* model, piRNA-mediated destabilization of transcripts in cytoplasm has been shown to provide posttranscriptional retrotransposon silencing in the germline [5]. Although evidence of the piRNA-mediated transcriptional silencing has been obtained [6, 7], the mechanism of the piRNA-mediated chromatin modification remains unknown. In this case, application of chromatin immunoprecipitation techniques which exploit specific antibodies is likely to be of low

Mikiko C. Siomi (ed.), *PIWI-Interacting RNAs: Methods and Protocols*, Methods in Molecular Biology, vol. 1093,
DOI 10.1007/978-1-62703-694-8_12, © Springer Science+Business Media, LLC 2014

efficiency for characterization of the chromatin status of the piRNA target loci. The assessment of the transcription rate may be more informative for this purpose.

A steady-state RNA level in the sample reflects additive effects of the RNA stabilization/degradation process and transcriptional rate. Nuclear run-on technique allows one to directly assess the level of nascent transcripts which, in its turn, is determined by the density of active RNA polymerase complexes over the genomic locus. In this method, isolated and washed nuclei are incubated in the presence of labeled nucleoside triphosphates. Previously engaged polymerases restart elongation, resulting in the accumulation of the labeled nascent transcripts whose abundance may be estimated by different methods, such as blot or microarray hybridization, RT-PCR, or deep sequencing.

Drosophila ovary is a mixture of somatic and germinal cells with different expression profile, which makes nuclear run-on experiments more difficult than those with homogenous cell cultures which are traditionally used for this assay. Despite the fact that sarkosyl is usually added to the reaction mixture to prevent transcription reinitiation events several researches indicate that initiation of transcription in the absence of sarkosyl is also very unlikely [8, 9]. Because of the high ploidy DNA index of the ovarian cells the presence of sarkosyl in the run-on reaction is undesirable as it results in the nucleus lysis and appearance of a high viscosity of the solution which may impede buffer component diffusion. According to this protocol, NRO assay is carried out in the absence of sarkosyl. Experimental evidence confirmed reliability and reproducibility of the NRO results obtained under the conditions chosen in this protocol [7, 9]. This approach combines advantages of different NRO techniques [7, 9, 10] and may be applied to estimate the transcription rate of any genomic loci of interest in *Drosophila* ovaries.

2 Materials

2.1 Nuclear Run-On Reaction

1. Dissected *Drosophila* ovaries from control and mutant lines, 100–300 ovary pairs per NRO assay (*see* **Note 1**).
2. PBS (phosphate-buffered saline) (pH 7.4).
3. Dounce homogenizer, 2 ml, pestle B (Sigma).
4. 100 mM dithiothreitol (DTT): store at −20 °C.
5. Complete protease inhibitor (Roche): 1,000× solution should be stored at −20 °C as 10 μl aliquots.
6. Buffer HB-A: 15 mM HEPES (pH 7.5), 10 mM KCl, 2.5 mM $MgCl_2$, 0.1 mM EDTA, 0.5 mM EGTA, 0.05 % NP40, 0.35 M sucrose, 1 mM DTT, 1×-recommended concentration complete

protease inhibitor (Roche). Store at –20 °C as 0.6 ml aliquots. DTT and protease inhibitor (*see* **items 4** and **5** in Subheading 2.1) are added to aliquots before use.

7. Buffer HB-B: 15 mM HEPES (pH 7.5), 10 mM KCl, 2.5 mM $MgCl_2$, 0.1 mM EDTA, 0.5 mM EGTA, 0.05 % NP40, 0.8 M sucrose, 1 mM DTT, 1× complete protease inhibitor (Roche). Store at –20 °C as 0.8 ml aliquots. DTT and protease inhibitor are added before use.
8. Buffer RB: 5 mM Tris–HCl (pH 8.0), 5 mM $MgCl_2$, 150 mM KCl, 1 mM DTT, 1× complete protease inhibitor (Roche). Store at –20 °C as 0.6 ml aliquots. DTT and protease inhibitor are added before use.
9. Miracloth membrane (Calbiochem).
10. 40 U/μl RNasin (Promega).
11. rNTP mix: 10 mM each rATP, rCTP, rGTP (Promega). Store at –20 °C as 10 μl aliquots.
12. 10 mM 5-bromouridine 5′-triphosphate solution in water (BrUTP, Sigma). Store at –20 °C as 10 μl aliquots.
13. Trizol-LS reagent (Invitrogen).
14. RNAsecure Resuspension solution (Ambion).
15. 96 % Ethanol.
16. 75 % Ethanol.
17. 15 mg/ml GlycoBlue (Ambion).

2.2 Immunopurification of NRO RNA Fraction

1. IP buffer: 150 mM NaCl, 50 mM Tris–HCl (pH 8.0), 0.05 % NP40, 1 mM EDTA, 1× complete protease inhibitor. Prepare this buffer fresh each time.
2. 2 % polyvinylpyrrolidone (PVP, Sigma).
3. Ultrapure BSA (Ambion).
4. Dynabeads M-280, sheep anti-mouse IgG (Invitrogen).
5. DynaMag™-2 magnet (Invitrogen) for magnetic separation of Dynabeads.
6. Anti-bromodeoxyuridine antibodies (clone PRB-1, mouse IgG1, Millipore) (*see* **Note 5**). There are several possibilities available for the immunopurification of BrUTP RNA. The NRO RNA fraction may be immunopurified using (1) the anti-bromodeoxyuridine agarose conjugate IgG (Santa Cruz Biotechnology); (2) mouse anti-bromodeoxyuridine antibodies and protein G sepharose; (3) anti-bromodeoxyuridine antibodies conjugated to biotin and streptavidin coupled Dynabeads (Invitrogen); (4) mouse anti-bromodeoxyuridine antibodies and Dynabeads sheep anti-mouse IgG (Invitrogen).
7. Trizol reagent (Invitrogen).

2.3 RT-qPCR (Reverse Transcription, Quantitative PCR)

1. SuperScript II reverse transcriptase (Invitrogen).
2. Random hexamer primer.
3. Gene-specific primers.

3 Methods

Principal steps of the NRO assay are represented in Fig. 1.

3.1 Isolation of Nuclei from Drosophila Ovarian Cells

The success of NRO assay is dependent mainly on the nuclei isolation step. All procedures and centrifugations are performed at 4 °C or on ice with ice-cold solutions. Homogenizer and tubes should be put on ice before procedure. The total time of the isolation of nuclei (from ovary homogenization to the beginning of the run-on reaction) should not exceed 1 h. Under these conditions, the transcription elongation by previously initiated RNA polymerases does not occur, which ensures a high yield of labeled run-on RNA during the following run-on reaction.

1. Dissect ovaries from ~100 to 300 females in PBS and store in a 1.5-ml tube on ice during isolation (up to 2 h).

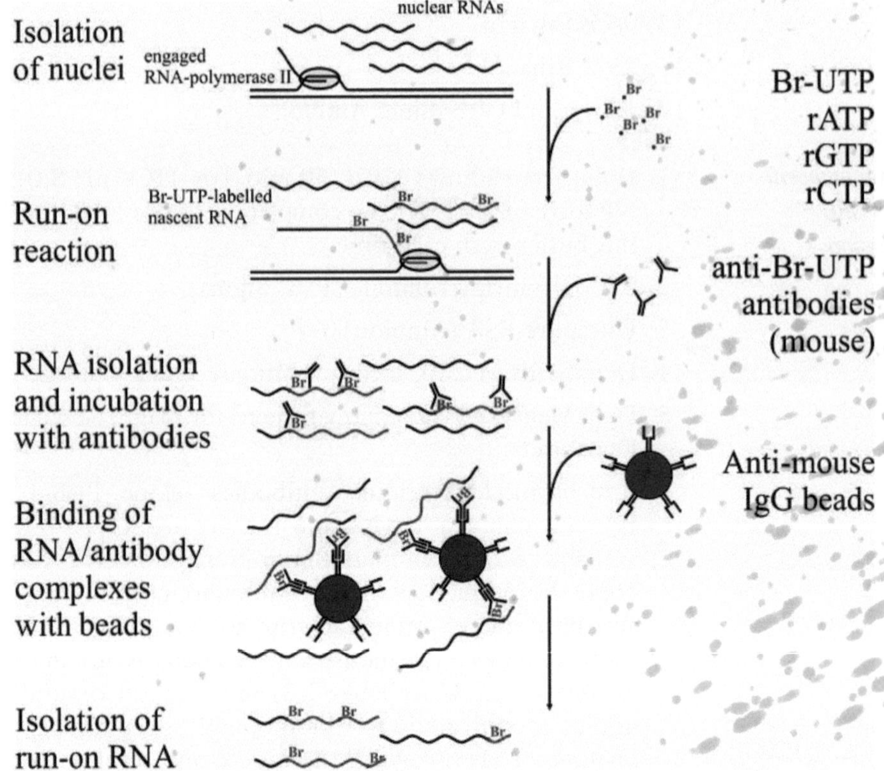

Fig. 1 Overview of the NRO assay. Schematic representation of the steps of the run-on transcription experiment using BrUTP and immunopurification of the BrUTP-labeled RNA fraction

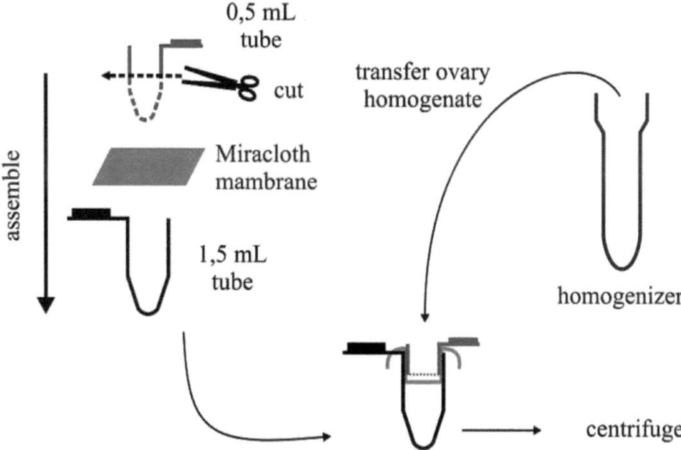

Fig. 2 Assembly of the Miracloth filter used for filtration of the ovary homogenate

2. Prepare the Miracloth filter. For this, put a 1-layer piece of membrane between a 0.5-ml tube with a cut-off bottom and a 1.5-ml tube (Fig. 2). Put it on ice.
3. Prepare a sucrose cushion: take a 1.5-ml tube, containing 0.8 ml of HB-B buffer as bottom phase and 0.3 ml of HB-A as upper phase.
4. Remove the PBS solution from the tube containing dissected ovaries, add 300 ml of HB-A buffer, transfer to the ice-cold Dounce homogenizer using a 1-ml tip and homogenize with 20–25 strokes of a tight-fitting pestle B holding the homogenizer on ice.
5. Filter the homogenate through Miracloth at 1,000 ×*g* for 2 min at 4 °C.
6. Discard a 0.5-ml tube and the Miracloth filter, carefully pipette the filtrate by a 1 ml tip and overlay it on the sucrose cushion.
7. Centrifuge at 10,000 ×*g* for 10 min in the Eppendorf centrifuge at 4 °C.
8. Remove the supernatant by a tip and resuspend the nucleus pellet in 0.5 ml of RB buffer using a 200-μl tip. Washing the nuclei in RB buffer prior to the run-on reaction is done to remove HB buffer components and an endogenous pool of triphosphates.
9. Centrifuge at 6,000 ×*g* for 7 min at 4 °C.
10. Remove the supernatant by a tip, being careful to avoid the nucleus pellet. Try to remove the upper fat layer, although its traces are not likely to interfere with the proper run-on reaction. Immediately proceed to Subheading 3.2.

3.2 NRO Reaction

This step includes the run-on reaction and isolation of the total nuclear RNA containing BrUTP-labeled run-on RNA.

1. Prepare 100 μl per probe of the core RB solution containing 0.5 mM of each rATP, rGTP, rCTP, BrUTP, and 1 U/μl of RNasin (*see* **Note 2**).
2. Resuspend the nucleus pellet in 100 μl of the RB mixture and incubate at 25 °C for 30 min (*see* **Note 3**).
3. Add three volumes (300 μl) of the Trizol-LS reagent per sample, vortex for 30 s, let the tube sit at room temperature (RT) for 3 min.
4. Add 1 volume (100 μl) of chloroform, vortex for 30 s, and leave at RT for 5 min.
5. Centrifuge at 13,000 × *g* for 10 min at 4 °C.
6. Transfer the top aqueous layer to a 1.5-ml tube, add 2 μl of GlycoBlue, 2.5 volumes of 96 % ethanol and incubate overnight (for at least 30 min) at −20 °C.
7. Centrifuge at 16,000 × *g* for 30 min at 4 °C. Discard the supernatant and wash the RNA pellet with 500 μl of 75 % ethanol.
8. Centrifuge at 16,000 × *g* for 5 min at 4 °C.
9. Discard the supernatant, air dry the pellet at RT for 5–10 min.
10. Add 30 μl of the RNAsecure Resuspension solution preheated at 60 °C, vortex briefly, heat at 60 °C for 10 min and vortex again.
11. Proceed with BrUTP RNA immunopurification or store the nuclear RNA at −20 °C.

3.3 Immunopurification of NRO RNA Fraction

Enrichment and purity of the BrUTP-labeled RNA immunopurified by different methods were estimated in the preliminary experiments using in vitro synthesized firefly and Renilla luciferase mRNAs. Better results were obtained when mouse anti-bromodeoxyuridine antibodies with subsequent binding to Dynabeads sheep anti-mouse IgG (Invitrogen) were used. The output of BrUTP RNA was 80 % relative to the input. The purified RNA fraction contained 98 % of the BrUTP-labeled RNA after a single selection. Therefore, NRO RNA is highly enriched for BrUTP-labeled RNAs as compared to contaminant RNAs, providing a desired specificity of this method. Therefore, we describe here the details of the immunoprecipitation protocol using mouse anti-bromodeoxyuridine antibodies and sheep anti-mouse IgG Dynabeads (Invitrogen).

1. Take 40 μl of the bead suspension and wash in 1 ml of IP buffer by rotating for 5 min at RT.
2. Separate the beads by the magnet, aspirate the supernatant, and add 300 μl of the core IP containing 0.1 % polyvinylpyrrolidone, 1 mg/ml ultrapure BSA, and 0.1 U/μl RNasin. Incubate at RT for 1 h while rotating.

3. While the beads are being blocked, perform binding of the RNA with antibodies (*see* **Note 4**). For this, add 29 μl of the total nuclear RNA from **step 11** in Subheading 3.2 to 250 μl of IP containing 1 U/μl RNasin and 5 μl (1 μg) of anti-bromodeoxyuridine antibodies (*see* **Note 5**). Incubate the RNA with antibodies at 4 °C for 1 h with rotation. Store 1 μl of the nuclear RNA at −20 °C for the analysis of the input, if desired (*see* **Notes 6** and 7).
4. Pellet the beads with a magnet, aspirate the blocking solution, and transfer the RNA–antibodies mixture to the beads. Incubate at 4 °C for 30 min with rotation.
5. Remove the solution and freeze an aliquot for analysis of the unbound fraction, if desired.
6. Wash the beads 5 times with 0.5 ml of the IP buffer containing 0.1 U/μl RNasin at 4 °C for 5–7 min per wash.
7. After the final wash, remove the wash solution and add 200 μl of Trizol reagent directly to the beads.
8. Vortex for 30 s. Let the tube sit for 3 min at RT.
9. Add 50 μl of chloroform, vortex 30 s, and incubate 5 min at RT.
10. Spin at 13,000 × *g* for 10 min at 4 °C.
11. Transfer the top aqueous layer to a 1.5-ml tube, add 1.5 μl of GlycoBlue, 2.5 volume of 96 % ethanol, and incubate overnight at −20 °C.
12. Centrifuge at 16,000 × *g* for 30 min at 4 °C. Aspirate the supernatant carefully and wash the RNA pellet with 300 μl of 75 % ethanol.
13. Centrifuge at 16,000 × *g* for 5 min at 4 °C.
14. Discard the supernatant carefully, air dry the pellet at RT for 5 min.
15. Add 15 μl of the RNAsecure Resuspension solution preheated at 60 °C, vortex briefly, heat at 60 °C for 10 min, and vortex again.
16. Store the NRO RNA at −20 °C.

3.4 Analysis of NRO RNA

Depending on the aim of the experiment, NRO RNA may be sequenced or hybridized to blots or cDNA microarrays. We use the ovarian NRO RNA to compare the transcriptional rate of retrotransposons in hetero- and homozygous piRNA pathway gene mutants using RT-qPCR [7]. We have chosen gene-specific primers to amplify ~100 bp of the retrotransposon 5′ UTR or ORF regions. It is important to choose those primers which correspond to the most conserved region of a retrotransposon. We will not describe in this Protocol the details of the RT-qPCR analysis which include standard steps.

1. Perform a reverse transcription reaction with SuperScript II according to the manufacturer's guidelines using the random hexamer primer. We take 5 μl of NRO RNA per reaction.
2. Perform the quantitative RT-PCR analysis of the genes of interest. The RT sample from **step 1** in Subheading 3.4 is enough to analyze about ten genes in duplicates (*see* **Notes 8** and **9**).

4 Notes

1. The amount of ovaries required for the NRO reaction depends on the peculiarities of the ovary development in particular *Drosophila* stocks or mutant lines. Usually, 2–3-day-old females are suitable for the ovary dissection. It is important that the ovaries of the *Drosophila* lines to be compared should be morphologically similar. Feeding flies with yeast improves the yield of ovaries.
2. Sarkosyl is traditionally added to the nuclear run-on reaction to prevent new transcription initiation events [11]. We perform the NRO assay in the absence of sarkosyl. Ovarian cells have the genome of a high ploidy index (up to 1,024 haploid DNA value). Addition of sarkosyl results in the nucleus lysis and appearance of a high viscosity of the solution. Decrease of the starting amount of material results in a low amount of the run-on RNA fraction. We have shown that the presence/absence of sarkosyl do not affect the result of the NRO assay [7]. Really, the event of transcription reinitiation in isolated and washed nuclei is very unlikely even in the absence of sarkosyl owing to the high energy requirements of promoter DNA unwinding.
3. In the run-on experiments performed in Zamore's lab [9], the amount of run-on RNA was shown to reach its maximum by the 30-th min. Under the conditions of our assay, the dynamic curves of run-on transcript accumulation were similar [7]. Thus, to obtain the amount of the run-on RNA sufficient for subsequent RT-PCR, we keep the run-on reaction for 30 min.
4. A higher yield of the NRO RNA fraction is observed when antibodies are incubated at first with the RNA and then RNA–antibody complexes bind to beads. Most likely, it depends on the peculiarities of the sterical specificity of antigene/antibody and antibody/bead complexes.
5. In our preliminary experiments, we found that BrUTP RNA-anti-bromodeoxyuridine antibody (clone PRB-1, biotin-conjugated mouse IgG_1, Millipore) complexes were more efficiently bound to the Dynabeads sheep anti-mouse IgG (Invitrogen) than to streptavidin coupled or protein G Dynabeads (Invitrogen). Therefore, we use this strategy for

immunopurification. We have found anti-BrdUTP antibodies, clone PRB-1, to be more efficient in the NRO assay than anti-BrdUTP, clone IIB5. We have not tested other commercially available mouse anti-BrdUTP antibodies.

6. Addition of the "cold" and BrUTP-labeled heterogeneous RNAs as tracers to each sample of the nuclear RNA before immunopurification may be useful to estimate the yield and purity of the NRO RNA fraction using RT-qPCR. For this, we add 0.2 ng of each of in vitro synthesized "cold" firefly and BrUTP-labeled Renilla RNAs per sample.
7. It is important to ensure that such processes as co-transcriptional RNA degradation or processing will not impact the amount of NRO RNA. Enrichment of NRO libraries by intronic sequences and highly unstable transcripts indicates that processing and RNA degradation in isolated nuclei are inhibited [10]. We have also shown that splicing is not effective after isolation of nuclei as the majority of *rp49* transcripts in the NRO RNA fraction are nonprocessed, while in the total nuclear RNA, the amount of unspliced *rp49* transcripts is ~60 times lower than that of the processed ones (Fig. 3). Enrichment of the NRO RNA in the unspliced transcripts may serve as a good control for the NRO RNA reliability. We have also demonstrated similar rates of

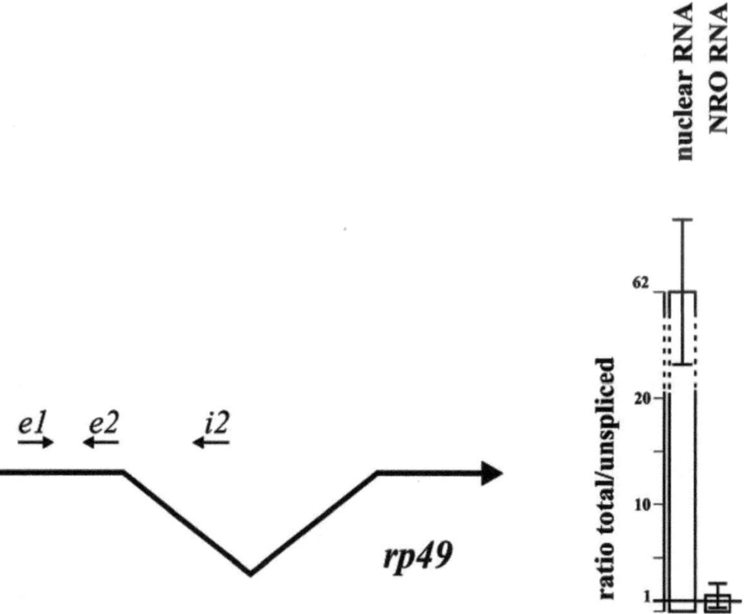

Fig. 3 The RT-qPCR analysis of *rp49* transcripts in the total nuclear RNA and in the NRO RNA fraction demonstrating that splicing is not effective in the isolated nuclei. Primers e1/e2 detect both spliced and unspliced *rp49* transcripts, while e1/i2 primers reveal unspliced transcripts

accumulation of *rp49* (constitutively expressed gene) and *HeT-A* (retrotransposon, piRNA target) run-on transcripts in the control and *piwi* knockdown flies. This fact indicates that no piRNA-mediated posttranscriptional or co-transcriptional degradation of *HeT-A* transcripts occurs in isolated nuclei of the control line. Thus, the chosen conditions allow us to estimate the transcriptional rate at the locus of interest.

8. When analyzing repetitive sequences, it is important to normalize RT-qPCR values to the gene copy number which can be easily determined using the genomic DNA of the lines to be compared.
9. A typical result of the RT-qPCR analysis of total ovarian and NRO RNAs in the piRNA pathway mutants is shown in Fig. 4. For those retrotransposons which are expressed in both somatic and germ ovarian cells, transcriptional effects occurring in the mutant

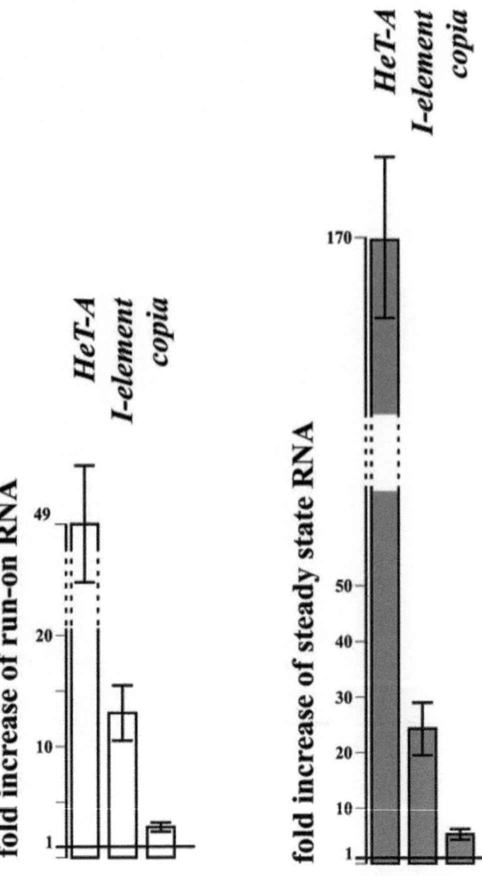

Fig. 4 The NRO analysis of the retrotransposon transcription in the ovaries of piRNA pathway gene *spn-E* mutants. The bars of histograms represent the ratio of retrotransposon to the *rp49* run-on transcript abundance in the ovaries of *spn-E*/*spn-E* related to this ratio in heterozygous *spn-E*/+ ovaries

germ line may be masked by the high transcriptional rate in follicular cells. In this case, the results of the NRO assay are difficult to interpret (for example, retrotransposon *copia*; *see* Fig. 4).

Acknowledgments

This work was supported by grants to A.K. from the Russian Academy of Sciences program for Molecular and Cell Biology and the Russian Foundation for Basic Researches (12-04-00797).

References

1. Aravin AA, Hannon GJ, Brennecke J (2007) The Piwi-piRNA pathway provides an adaptive defense in the transposon arms race. Science 318:761–764
2. Hamilton A, Voinnet O, Chappell L, Baulcombe D (2002) Two classes of short interfering RNA in RNA silencing. EMBO J 21:4671–4679
3. Liu Y, Mochizuki K, Gorovsky MA (2004) Histone H3 lysine 9 methylation is required for DNA elimination in developing macronuclei in Tetrahymena. Proc Natl Acad Sci U S A 101:1679–1684
4. Volpe TA, Kidner C, Hall IM, Teng G, Grewal SI, Martienssen RA (2002) Regulation of heterochromatic silencing and histone H3 lysine-9 methylation by RNAi. Science 297:1833–1837
5. Lim AK, Tao L, Kai T (2009) piRNAs mediate posttranscriptional retroelement silencing and localization to pi-bodies in the Drosophila germline. J Cell Biol 186:333–342
6. Klenov MS, Lavrov SA, Stolyarenko AD, Ryazansky SS, Aravin AA, Tuschl T, Gvozdev VA (2007) Repeat-associated siRNAs cause chromatin silencing of retrotransposons in the Drosophila melanogaster germline. Nucleic Acids Res 35:5430–5438
7. Shpiz S, Olovnikov I, Sergeeva A, Lavrov S, Abramov Y, Savitsky M, Kalmykova A (2011) Mechanism of the piRNA-mediated silencing of Drosophila telomeric retrotransposons. Nucleic Acids Res 39:8703–8711
8. Jackson DA, Iborra FJ, Manders EM, Cook PR (1998) Numbers and organization of RNA polymerases, nascent transcripts, and transcription units in HeLa nuclei. Mol Biol Cell 9:1523–1536
9. Sigova A, Vagin V, Zamore PD (2006) Measuring the rates of transcriptional elongation in the female Drosophila melanogaster germ line by nuclear run-on. Cold Spring Harb Symp Quant Biol 71:335–341
10. Core LJ, Waterfall JJ, Lis JT (2008) Nascent RNA sequencing reveals widespread pausing and divergent initiation at human promoters. Science 322:1845–1848
11. Rougvie AE, Lis JT (1988) The RNA polymerase II molecule at the 5′ end of the uninduced hsp70 gene of D. melanogaster is transcriptionally engaged. Cell 54:795–804

Chapter 13

Combined RNA/DNA Fluorescence In Situ Hybridization on Whole-Mount *Drosophila* Ovaries

Sergey Shpiz, Sergey Lavrov, and Alla Kalmykova

Abstract

DNA FISH (fluorescent in situ hybridization) analysis reveals the chromosomal location of the gene of interest. RNA in situ hybridization is used to examine the amounts and cell location of transcripts. This method is commonly used to describe the localization of processed transcripts in different tissues or cell lines. Gene activation studies are often aimed at determining the mechanism of this activation (transcriptional or posttranscriptional). Elucidation of the mechanism of piRNA-mediated silencing of genomic repeats is at the cutting edge of small RNA research. The RNA/DNA FISH technique is a powerful method for assessing transcriptional changes at any particular genomic locus. Colocalization of the RNA and DNA FISH signals allows a determination of the accumulation of nascent transcripts at the transcribed genomic locus. This would be suggest that this gene is activated at the transcriptional (or co-transcriptional) level. Moreover, this method allows for the identification of transcriptional derepression of a distinct copy (copies) among a genomic repeat family. Here, a RNA/DNA FISH protocol is presented for the simultaneous detection of RNA and DNA in situ on whole-mount *Drosophila* ovaries using tyramide signal amplification. With subsequent immunostaining of chromatin components, this protocol can be easily extended for studying the interdependence between chromatin changes at genomic loci and their transcriptional activity.

Key words FISH, Transposable elements, *Drosophila*, Ovaries, Transcription, TSA, piRNAs

1 Introduction

Expression of transposable elements (TEs) in *Drosophila* germline cells is suppressed by piRNAs (PIWI-interacting RNAs) [1, 2]. Disruption of the piRNA machinery leads to activation of transposable element expression. Most likely, both posttranscriptional and transcriptional pathways may be involved in the piRNA-mediated silencing of the same target gene. TEs are repressed via piRNA assistance, at least partly, at the transcriptional level [3, 4]. Accumulation of nascent transcripts may be visualized in the proximity of the activated genomic loci by RNA/DNA fluorescent in situ hybridization (FISH, [3, 5, 6]). Expression of different copies

Mikiko C. Siomi (ed.), *PIWI-Interacting RNAs: Methods and Protocols*, Methods in Molecular Biology, vol. 1093,
DOI 10.1007/978-1-62703-694-8_13,

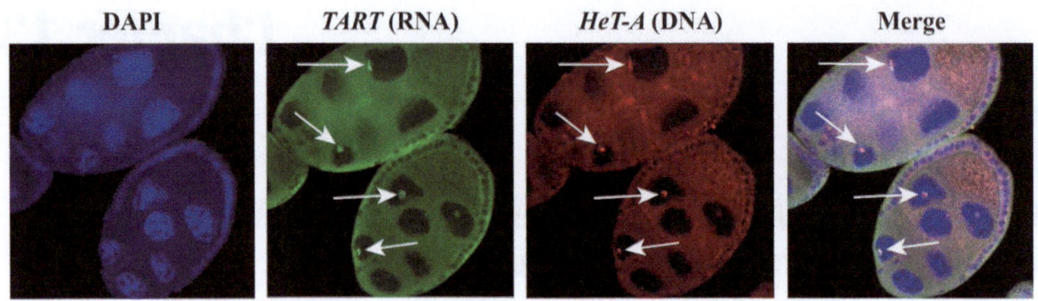

Fig. 1 Telomeric retroelement *TART* antisense transcripts accumulate in telomeric regions of the nurse cell nuclei in homozygous *spn-E* mutants. Antisense TART transcripts (*green*) are colocalized with the HeT-A DNA probe (*red*), which serves here as a telomere marker. Thus, these data show the accumulation of nascent TART transcripts at the site of transcription

of the same multiple copy TE family may be regulated in different ways. The position of a particular TE genomic copy may be marked using a DNA FISH probe specific to the unique genomic regions flanking this copy. Such an approach may be useful to detect transcript accumulation for the exact TE copy (or a local set of TE copies). Subsequent chromatin protein immunostaining of the same sample can reveal a correlation between the transcriptional activity of a particular TE copy and the chromatin status of the corresponding genomic locus.

Drosophila ovaries are a mixture of somatic and germinal cells with different gene expression profiles. Conventional techniques, like RT-PCR or chromatin immunoprecipitation (ChIP), cannot be easily applied to study gene expression changes on a per-cell basis in such heterogeneous tissues like ovaries. Moreover, knowledge about the expression of a particular copy of TE family in a specific cell type may be very informative. This can be achieved using the DNA/RNA FISH technique.

Here, we describe a protocol in which RNA in situ hybridization with a single-stranded riboprobe is performed first because RNA is less stable in comparison with DNA. Next, DNA FISH is done according to standard protocols [7]. The procedure of DNA FISH may impair the level of RNA FISH signals, so we use tyramide signal amplification (TSA, [8, 9]) to preserve RNA signals in the combined DNA/RNA FISH protocol. We applied this technique to describe the transcription of the telomeric TEs *HeT-A* and *TART* in the ovaries of the piRNA pathway mutants [3]. TE transcripts are undetectable in wild-type and heterozygous mutant flies. However, we observed an accumulation of telomeric retrotransposon nascent transcripts at the telomeres in homozygous *spn-E* mutant flies; this suggests the transcriptional activation of the telomeric retrotransposons in the piRNA mutants (Fig. 1).

2 Materials

1. 1× PBS (phosphate-buffered saline) (pH 7.4): 137 mM NaCl, 2.7 mM KCl, 10 mM Na_2HPO_4, 1.8 mM KH_2PO_4, adjust to pH 7.4 with HCl.
2. 1× PBT: 1× PBS containing 0.1 % Tween 20.
3. 3.7 % formaldehyde in 1× PBT.
4. 20× SSC (saline-sodium citrate): 3 M NaCl, 0.3 M Na-citrate, adjust to pH 7.0 with HCl.
5. Hybridization buffer: 50 % formamide, 5× SSC, 0.1 % Tween 20, 100 μg/ml denatured DNA (from herring sperm), 50 μg/ml heparin.
6. Triton X-100.
7. Ultrapure BSA (Ambion).
8. Fluorophore-conjugated tyramide.
9. 3 % H_2O_2.
10. Anti-DIG-POD antibodies (Roche).
11. Fluorophore-conjugated streptavidin.
12. Normal goat serum (NGS).
13. Formamide deionized.
14. CHAPS (Sigma).
15. BioNick™ Labeling System (Invitrogen).
16. DIG RNA-labeling mix (Roche).
17. TranscriptAid kit (Fermentas).
18. Bio-Rad Micro Bio-Spin 30 column.
19. Image-iT FX (Invitrogen).
20. Slow Fade Gold (Invitrogen).

3 Methods

The key point of this protocol is tyramide signal amplification (TSA). The mechanism of TSA is shown in Fig. 2. TSA is an enzyme-mediated detection method that utilizes the catalytic activity of horseradish peroxidase (HRP). In the first step, DIG or a Biotin-labeled probe binds via hybridization followed by secondary detection of the probe with an HRP-conjugated antibody or streptavidin conjugate, respectively. Next, HRP activates multiple copies of dye-labeled tyramide in the presence of hydrogen peroxide resulting in highly reactive, short-lived tyramide radicals covalently coupled to tyrosine residues in the proximity of the HRP-target interaction site. Some TSA parameters were optimized.

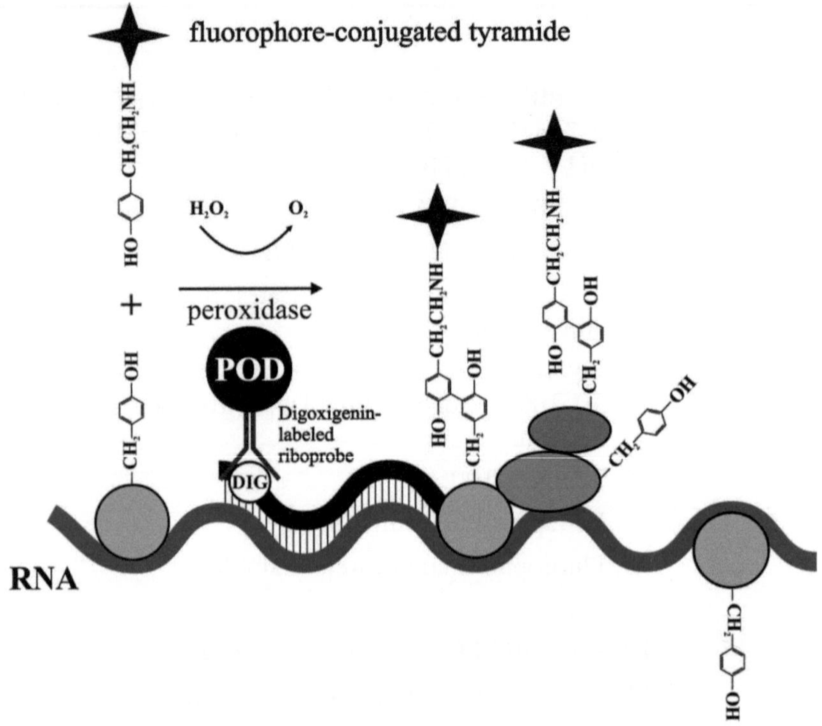

Fig. 2 Mechanism of tyramide signal amplification (TSA)

Tyramide concentration. A high tyramide concentration accelerates its binding but leads to higher nonspecific backgrounds and to consumption of the reagent. We searched for an optimal tyramide concentration within a range of 0.2–8 μg/ml and found it to be 4 μg/ml.

Peroxide concentration. Hydrogen peroxide concentration determines the rates of tyramide binding and peroxidase inactivation. In a range of 0.001–0.02 %, we found a 0.012 % concentration of peroxide to be optimal.

Duration of amplification step. The level of signal amplification depends on the time of reaction: the longer the incubation, the more fluorescent molecules are bound in the hybridization zone. Signal amplification rate is limited by two factors, a gradual increase in background levels and inactivation of probe-bound peroxidase (partly via binding to tyramide). According to previously published protocols, incubation with tyramide varied from 5 up to 45 min. In a range of 15–90 min, we found that 30–60 min was optimal (depending on the intensity of the primary signal).

Dye type. Combinations of different RNA-labeling methods and detection methods have been compared: (1) DIG-labeled probe—DIG antibody conjugated with peroxidase [anti-DIG-POD (Roche 11207733910)]; (2) biotinylated probe—avidin conjugated with peroxidase [avidin-HRP conjugate (Sigma A7419)]. In both cases, an optimal peroxidase conjugate dilution was found to be 1:500. However, when we used biotinylated probes, background signals were higher than in the case of DIG probes. Biotinylated probes appear to be useful for the detection of mRNA of highly expressed genes. In these cases, the hybridization signal is visible even with high background levels. Generally, we used DIG-labeled RNA probes.

3.1 Preparation of Probes

1. Riboprobe preparation: Carry out in vitro transcription (IVT) using DIG RNA labelling mix (Roche) and the TranscriptAid kit (Fermentas) or its analogues according to the manufacturer's instructions. PCR-amplified fragments containing the T7 RNA polymerase promoter, or DNA fragments cloned in plasmids containing RNA polymerase promoters, may be used as templates for the synthesis of riboprobes. Synthesized RNA is purified using Bio-Rad Micro Bio-Spin 30 columns according to manufacturer's specifications. Dry out eluates in a Speedvac up to ~10 μl, and then add 1 ml of hybridization buffer containing 2 μg/μl of yeast RNA. The final concentration of DIG-labeled RNA in this solution is ~10 ng/μl. Alternatively, RNA obtained in the IVT reaction may be precipitated using 5 M ammonium acetate and ethanol. The RNA probe stock solution can be stored at −20 °C for 6 months.
2. DNA probe preparation: Use the BioNick™ Labeling System (Invitrogen) according to manufacturer's instructions to prepare the biotin-labeled probes. Dissolve labeled probe in 80 μl of hybridization buffer containing 2 μg/μl of yeast RNA. The final concentration of biotinylated probe in the stock solution is ~25 ng/μl. The DNA probe stock solution can be stored at −20 °C for 1 year.

3.2 RNA FISH

1. Dissect the ovaries from ~50 *Drosophila* females in PBT and store in a 1.5-ml Eppendorf tube on ice during isolation (up to 2 h). For RNA FISH, it is important to preserve organ integrity at this step.
2. Fix ovaries for 20 min in 3.7 % formaldehyde, 1× PBT at room temperature (RT) on a rotating wheel.
3. Wash ovaries with PBT for 5 min three times on a rotating wheel.
   After this step, tissues can be stored at 4 °C overnight.
   It is recommended to disunite the ovarioles with needles or tweezers (*see* **Note 1**).
4. Incubate ovaries in 0.6 % Triton X-100, 1× PBS for 10 min at RT on a rotating wheel (*see* **Note 2**).
5. Wash ovaries with PBT for 5 min three times on a rotating wheel. After this step, tissues can be stored at 4 °C overnight.

6. Prepare fresh solutions of 20, 50, and 80 % hybridization buffer in PBT. Transfer samples stepwise to hybridization buffer by incubation in 1 ml of each solution for 15 min.
7. Incubate samples in 100 % hybridization buffer for 15 min.
8. Preincubate ovaries in hybridization buffer at 42 °C for 1–3 h.
9. Denature 100 μl of DIG-labeled riboprobe solution from **step 1** in Subheading 3.1 at 80 °C for 10 min.
10. After removing the supernatant from **step 8**, add 100 μl of denaturated DIG-labeled riboprobe solution from **step 9**. Overlay the sample with mineral oil.
11. Incubate at 42 °C overnight with shaking (*see* **Note 3**).
12. Wash two times with 1 ml of 50 % formamide, 2× SSC, 0.3 % CHAPS for 15 min at 60 °C with shaking.
13. Wash with 1 ml of 40 % formamide, 2× SSC, 0.3 % CHAPS for 15 min at 42 °C with shaking.
14. Incubate sequentially for 15 min at 42 °C with shaking in 1 ml of the following solutions:
    (a) 1× PBT, 30 % formamide
    (b) 1× PBT, 20 % formamide
    (c) 1× PBT, 10 % formamide
15. Wash ovaries with PBT for 5 min three times at RT.
16. Block in PBT containing 5 % NGS, 1 % BSA, 0.3 % Triton X-100 for 1–3 h.
17. Optional. Incubate with 1 % H_2O_2 in PBT for 30 min at RT to inactivate endogenous peroxidase (*see* **Note 4**).
18. Incubate with anti-DIG POD antibodies (Roche) (diluted at 1:500) in PBT, 0.3 % Triton X-100, 3 % NGS for at least 1 h on a rotating wheel at RT.
19. Wash ovaries with PBT for 10 min five times at RT.
20. Incubate with fluorophore-conjugated tyramide (4 μg/ml) in PBT for 30 min.
21. Add H_2O_2 to a final concentration of 0.012 % and incubate for 30–60 min at RT with rotation.
22. Wash ovaries with PBT for 5 min three times at RT.
23. Check RNA signal under fluorescent microscope (*see* **Note 5**). Tissues can be stored at 4 °C for several days after this step.

3.3 DNA FISH

Take tissue samples from **step 22** in Subheading 3.2.

1. Optional. Add RNAse A to a final concentration of 100 μg/ml. Incubate for 4 h at 37 °C or overnight at 4 °C.
2. Wash ovaries with PBT for 5 min three times at RT.

3. Prepare fresh solutions of 20, 50, and 80 % hybridization buffer in PBT. Incubate samples for 15 min in 1 ml of each solution.
4. Incubate in 100 % hybridization buffer for 15 min to replace the PBT with hybridization buffer.
5. After removing the supernatant from the last step, add 20 μl of biotin-labeled probe solution from **step 2** in Subheading 3.1. If you use biotinylated oligonucleotides as a probe, their concentration in hybridization solution should be 0.5 pM/μl. Overlay with mineral oil.
6. Incubate ovaries at 80 °C for 15 min to denature DNA.
7. Hybridize with probe overnight at 37 °C (*see* **Note 6**) with shaking (*see* **Note 3**).
8. Wash two times with 1 ml of 50 % formamide, 2× SSC, 0.3 % CHAPS for 15 min at 37 °C with shaking.
9. Wash with 1 ml of 40 % formamide, 2× SSC, 0.3 % CHAPS for 15 min at 37 °C with shaking.
10. Incubate sequentially for 15 min at 37 °C with shaking in 1 ml of the following solutions:
    (a) 1× PBT, 30 % formamide
    (b) 1× PBT, 20 % formamide
    (c) 1× PBT, 10 % formamide
11. Wash ovaries with PBT for 5 min three times at RT. Tissues can be stored at 4 °C for several days after this step.

3.4 DNA Staining and Immunostaining

Take tissue samples from **step 11** in Subheading 3.3.

1. Replace PBT with Image-iT FX (Invitrogen) and incubate for 30 min at RT (*see* **Note 7**).
2. Wash ovaries with PBT for 5 min three times at RT.
3. Optional: Block in PBT containing 5 % NGS, 1 % BSA, 0.3 % Triton X-100 for 1–3 h.
4. Incubate with primary antibodies specific for the protein of interest in PBT containing 3 % NGS for 1 h on a rotating wheel.
5. Wash ovaries with PBT for 5 min three times at RT.
6. Incubate with fluorophore-conjugated streptavidin and fluorophore-conjugated secondary antibodies in PBT, 0.3 % Triton X-100, 3 % NGS for 1 h on a rotating wheel (*see* **Note 7**).
7. Wash ovaries with PBT, 0.3 % Triton X-100 for 5 min three times at RT.
8. Incubate in 5 μM DAPI in PBT for 10 min at RT.
9. Wash ovaries in PBT for 5 min three times at RT.
10. Remove supernatant and add 60 μl of Slow Fade Gold solution (Invitrogen).

11. Pipette ovaries onto a slide. Lower a coverslip onto the tissues. Seal the coverslip to the slide with nail polish.
12. Analyze samples by conventional fluorescence or confocal microscopy.

4 Notes

1. The disjunction of distinct ovarioles is recommended at this step to enhance the accessibility of ovarian cells by the labeled probes and antibodies.
2. Incubation in 0.6 % Triton X-100 is used to improve subsequent probe and antibody penetration. This step can be replaced by the following procedures: treatment with 50 μg/ml proteinase K solution for 2–15 min (should be adjusted); incubation in 0.3 % Triton X-100 with 0.3 % sodium deoxycholate in PBS twice for 30 min; freezing/thawing.
3. **Step 11** in Subheading 3.2, **steps 6** and 7 in Subheading 3.3, as well as washing steps, may be performed in an Eppendorf Thermo mixer at 850 rpm shaking. All solutions used in these steps should be preheated to the appropriate temperature.
4. The activity of endogenous peroxidases may result in a high level of background; therefore, it may be useful to inactivate them with a high concentration of hydrogen peroxide before binding. However, in formaldehyde-fixed samples, endogenous peroxidases are inactivated. We have not pretreated samples with hydrogen peroxide.
5. Hybridization conditions applied in Subheading 3.2 allowed us to detect RNA rather than genomic DNA. To exclude the possibility of RNA/DNA hybridization at this stage, an RNase H control may be done. After hybridization and washing (after **step 15** in Subheading 3.2), incubate ovaries in PBS containing 0.3 u/μl RNase H for 1 h at 37 °C.
6. If biotinylated oligonucleotides are used, the hybridization temperature should be adjusted taking into account the length and melting temperature of the oligonucleotide. Commonly, 30 °C is a suitable hybridization temperature.
7. Background staining emerging from fluorescent reagents is largely eliminated when the Image-iT™ FX signal enhancer is applied to fixed and permeabilized cells prior to staining. It may occur that, in the course of immunostaining with particular antibody combinations, incubation with Image-iT FX reduces signal-to-noise ratios; it is recommended that you check if it occurs in your specific case. If so, skip **steps 1** and **2** in Subheading 3.4, but perform **step 3** in Subheading 3.4 for blocking.

Acknowledgments

This work was supported by grants to S.S. from the Russian Foundation for Basic Researches (12-04-00996) and to A.K. from the Russian Academy of Sciences Program for Molecular and Cell Biology.

References

1. Aravin AA, Hannon GJ, Brennecke J (2007) The Piwi-piRNA pathway provides an adaptive defense in the transposon arms race. Science 318:761–764
2. Vagin VV, Sigova A, Li C, Seitz H, Gvozdev V, Zamore PD (2006) A distinct small RNA pathway silences selfish genetic elements in the germline. Science 313:320–324
3. Shpiz S, Olovnikov I, Sergeeva A, Lavrov S, Abramov Y, Savitsky M, Kalmykova A (2011) Mechanism of the piRNA-mediated silencing of Drosophila telomeric retrotransposons. Nucleic Acids Res 39:8703–8711
4. Klenov MS, Lavrov SA, Stolyarenko AD, Ryazansky SS, Aravin AA, Tuschl T, Gvozdev VA (2007) Repeat-associated siRNAs cause chromatin silencing of retrotransposons in the Drosophila melanogaster germline. Nucleic Acids Res 35:5430–5438
5. Takizawa T, Gudla PR, Guo L, Lockett S, Misteli T (2008) Allele-specific nuclear positioning of the monoallelically expressed astrocyte marker GFAP. Genes Dev 22:489–498
6. Shpiz S, Kwon D, Rozovsky Y, Kalmykova A (2009) rasiRNA pathway controls antisense expression of Drosophila telomeric retrotransposons in the nucleus. Nucleic Acids Res 37: 268–278
7. Lavrov S, Dejardin J, Cavalli G (2004) Combined immunostaining and FISH analysis of polytene chromosomes. Methods Mol Biol 247:289–303
8. Speel EJ, Komminoth P (1999) CARD in situ hybridization: sights and signals. Endocr Pathol 10:193–198
9. Speel EJ, Hopman AH, Komminoth P (1999) Amplification methods to increase the sensitivity of in situ hybridization: play card(s). J Histochem Cytochem 47:281–288

Chapter 14

Fast and Accurate Method to Purify Small Noncoding RNAs from Drosophila Ovaries

Thomas Grentzinger and Séverine Chambeyron

Abstract

The recent development of High Throughput Sequencing technology has boosted the study of small regulatory RNA populations. A critical step prior to cloning and sequencing is purification of small RNA populations. Here, we report the optimization of an anion-exchange chromatography procedure in order to purify small regulatory RNAs bound on proteins. We developed this procedure to make it less time-consuming since our improved method no longer requires specific equipment and can easily be performed at the bench. We believe that our procedure will increase the robustness and accuracy of small RNA libraries in the future.

Key words Small regulatory RNA purification, Anion-exchange chromatography, *Drosophila melanogaster*

1 Introduction

Transposable elements are usually silenced especially in germline where genome integrity is crucial for the maintenance of accurate genetic information. However, rare exceptions of loss of control of TE regulation in germline, called hybrid dysgenesis, were reported in Drosophila females. This phenomenon is observed in the progeny of crosses between male of strains containing multiple copies of a functional transposable element and female of strains devoid of the same copies. Thus high rates of transposition occurred in the F1 progeny resulting in various genetic abnormalities including mutations, chromosome rearrangements and, ultimately, the sterility of this F1 progeny [1–4].

The development of RNA-seq approaches allowed characterization of a new class of small regulatory noncoding RNAs, the piwi-interacting RNAs (piRNAs), involved in the genome defense against the TE mobility [5, 6]. The piRNAs, associated with the PIWI subfamily of Argonaute proteins, form a silencing effector complex called piwi-RNA Inducing Silencing Complex (piRISC). In the Drosophila ovaries, three PIWI proteins are expressed in the

Mikiko C. Siomi (ed.), *PIWI-Interacting RNAs: Methods and Protocols*, Methods in Molecular Biology, vol. 1093,
DOI 10.1007/978-1-62703-694-8_14,

germline [Piwi, Aubergine (Aub), and Argonaute 3 (Ago3)] and only one, Piwi, is expressed in the surrounding somatic tissues.

Discovery of this new regulatory pathway offered the opportunity to understand both *I*- and *P*-hybrid-dysgenesis mechanisms [7]. Small noncoding RNA population comparisons between strains containing multiple copies of functional I-element or P-element, strains devoid of them and the F1 progeny after dysgenic crosses revealed different contents of piRNAs targeting these elements [7, 8]. Indeed, strains containing multiple copies of I- or P-element accumulate high levels of piRNAs that are maternally deposited allowing an efficient silencing response in the F1 progeny, whereas strains devoid of them are unable to do it. Thus, the hybrid dysgenesis phenotype, observed in the F1 progeny resulting from the loss of the functional TE repression, is due to a lack of maternal piRNA inheritance.

While the improvement of the High Throughput Sequencing allowed obtention of more accurate views of small RNA populations of a given cell or tissue, the small RNA isolation procedure became a crucial stage for the generation of robust and relevant sequencing data.

In 2006, a novel strategy based on anion-exchange chromatography was proposed as an alternative to the standard size-isolation purification procedure. Briefly, the strategy is based on the adsorption and reversible binding of negatively charged biomolecules to a positively charged insoluble matrix. piRISC can thus be recovered using mild-salt concentration elution, whereas contaminant RNAs (such as the 30-nt length ribosomal 2S RNA (2S rRNAs) and degradation products) remain fixed on the column. This method has proved its worth to purify piRISC from rat testis [9], Drosophila cell line [10], and Drosophila ovaries [11] nevertheless it was still not routinely used since it relied on advanced technical skills, expensive material and was time-consuming.

In this chapter we present an adaptation of this chromatography procedure applicable to all laboratories allowing a manual isolation of piRISC-associated piRNAs from Drosophila ovaries at the bench.

2 Materials

2.1 Flies Care, Ovaries Dissection, and Native Lysis

1. Standard Drosophila medium: 11.25 g agar is solubilized in 500 mL of hot water. 83.3 g yeast, 83.3 g corn flour, 33 mM methyl-4-hydroxybenzoate (Applichem) solubilized in 50 mL ethanol are added to the mix. The volume is completed to 1 L with water. Medium is heated at 90 °C for 90 min and then poured in Drosophila vials.
2. PBS 0.1 %T: Phosphate-Buffered Saline (PBS), 0.1 % (v/v) Tween 20.

3. Lysis buffer (also named Low-salt HQ buffer): 20 mM HEPES–KOH (pH 7.9), 10 % (v/v) glycerol, 1.5 mM $MgCl_2$, 0.2 mM ethylene-diamine tetraacetic acid (EDTA), 1 mM dithiothreitol (DTT), 0.2 mM phenylmethylsulfonyl fluoride (PMSF), 100 mM potassium acetate (CH_3CO_2K). Solution is filtered through a 0.22 μm cartridge and degassed under vacuum, then stored at 4 °C.
4. Motor-driven polypropylene pestle for 1.5-mL microcentrifuge tube (Sigma-Aldrich).

2.2 Manual HQ Chromatography

1. HiTrap Q HP 1-mL column (GE Healthcare).
2. HQ regeneration buffer: 20 mM HEPES–KOH (pH 7.9), 10 % (v/v) glycerol, 1.5 mM $MgCl_2$, 0.2 mM EDTA, 1 mM DTT, 0.2 mM PMSF, 1 M CH_3CO_2K. Solution is filtered through a 0.22 μm cartridge and degassed under vacuum, then stored at 4 °C.
3. Low-salt HQ buffer: 20 mM HEPES–KOH (pH 7.9), 10 % (v/v) glycerol, 1.5 mM $MgCl_2$, 0.2 mM ethylene-diamine tetraacetic acid (EDTA), 1 mM dithiothreitol (DTT), 0.2 mM phenylmethylsulfonyl fluoride (PMSF), 100 mM CH_3CO_2K. Solution is filtered through a 0.22 μm cartridge and degassed under vacuum, then stored at 4 °C.
4. High-salt HQ buffer: 20 mM HEPES–KOH (pH 7.9), 10 % (v/v) glycerol, 1.5 mM $MgCl_2$, 0.2 mM EDTA, 1 mM DTT, 0.2 mM PMSF, 4 M CH_3CO_2K. Solution is filtered through a 0.22 μm cartridge and degassed under vacuum, then stored at 4 °C.
5. Column storage buffer: 20 % (v/v) absolute ethanol in bidistilled water, 0.02 % (w/v) sodium azide (NaN_3).

2.3 RNA Extraction from Pooled Eluted Fractions

1. 2-mL microcentrifuge RNase-free tubes.
2. 1.5 mL Homo-Polymer maximum recovery microcentrifuge tubes (Axygen).
3. Acidic phenol/chloroform (5:1) (pH 4.5) (Ambion/LIFE Technologies).
4. GlycoBlue (Ambion/LIFE Technologies).
5. 3 M sodium acetate (pH 5.5) (Ambion/LIFE Technologies).

2.4 Small RNA Polyacrylamide Gel Isolation

1. 1.5-mL Homo-Polymer maximum recovery microcentrifuge tubes.
2. 10 mM 18-nt long RNA ladder: 5′-CGUACGCG GUUUAAACGA-3′ (Integrated DNA Technologies).
3. 10 mM 30-nt long RNA ladder: 5′-CGUACGCGGAAUA GUGUUUAAACUAAUUGU-3′ (Integrated DNA Technologies).

4. 10,000 U/mL T4 Polynucleotide kinase accompanied with 10× PNK buffer (New England Biolabs).
5. 10 Ci/mmol, 2 mCi/mL ATP [γ-^{32}P] (PerkinElmer).
6. DEPC-treated water.
7. Illustra Microspin G-25 column (GE Healthcare).
8. 5× TBE (pH 8.0): 0.45 M Tris-borate, 1 mM EDTA. Store at room temperature.
9. 4× RNA loading blue: 80 % (v/v) formamide, 20 % (w/v) sucrose, 1× TBE, 0.025 % (w/v) xylene cyanol, 0.025 % (w/v) bromophenol blue. Mix is filtered through a 0.22 μm filter and stored at room temperature.
10. Acrylamide gel stock solution: 15 % acrylamide:bisacrylamide, 8 M urea, 1× TBE. To obtain this solution mix 210.2 g urea, 50 mL 5× TBE and 75 mL bidistilled water. Heat under agitation until urea is solubilized then add 187.5 mL 40 % 29:1 acrylamide/bisacrylamide (MP Biomedicals) and complete to 500 mL with bidistilled water. Solution is filtered through a 0.22 μm cartridge and degassed under vacuum, then stored at 4 °C in a bottle protected from light.
11. APS: 10 % (w/v) ammonium persulfate in DEPC-treated water. Store at −20 °C.
12. 5 M NaCl.
13. Spin X 0.45 μm cartridge (Costar/Corning Life Science).
14. GlycoBlue (Ambion/LIFE Technologies).
15. Recombinant RNasin Ribonuclease inhibitor (Promega).

2.5 HQ Small RNA Quality Control

1. 1.5-mL Homo-Polymer maximum recovery microcentrifuge tubes.
2. Kinase Max kit (Ambion/LIFE Technologies).
3. 10 Ci/mmol, 2 mCi/mL ATP [γ-^{32}P].
4. Illustra Microspin G-25 column.
5. 4× RNA loading blue: 80 % (v/v) formamide, 20 % (w/v) sucrose, 1× TBE, 0.025 % (w/v) xylene cyanol, 0.025 % (w/v) bromophenol blue. Mix is filtered through 0.22 μm filter and stored at room temperature.

3 Methods

Our manual procedure allows a fast and basic purification of protein-bound small noncoding RNAs using anion-exchange chromatography at the bench. A day-by-day timeline for the procedure called manual HQ procedure is provided in Fig. 1a.

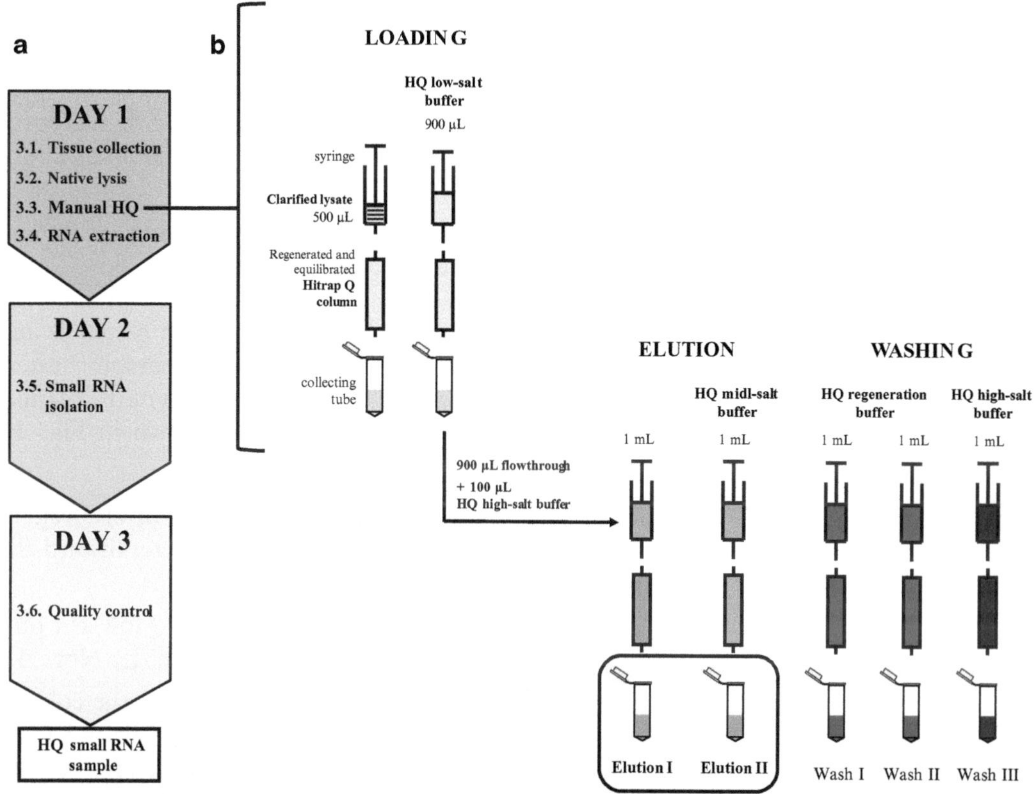

Fig. 1 Overview of the manual HQ small RNA extraction procedure. (**a**) Workflow illustrating a day-by-day outline for the basic steps of the manual HQ purification procedure. (**b**) Schematic representation of manual anion-exchange chromatography purification of small RNAs. Clarified lysate (*black line*) is manually loaded onto Hitrap Q HP column with 1 mL syringe. Elution is performed by increasing the ionic strength of the buffer. Gray level increases with ionic strength of the buffers. Elution fraction I and II are pooled prior to nucleic acid extraction. Washing the column is an optional step that allows its regeneration for reuse

3.1 Flies Care, Ovaries Dissection, and Native Lysis

1. Bred Drosophila stocks in standard Drosophila medium at 23 °C.
2. Transfer 1-day-old flies to medium complemented with fresh yeast for 3 days.
3. Dissect 20–50 pairs of ovaries on ice in cold PBS 0.1 %T and transfer them to 1.5-mL microcentrifuge tubes. Spin briefly on microcentrifuge and remove supernatant.
4. Wash tissues by adding 500 µL cold PBS 0.1%T, homogenize gently by tapping. Spin briefly on microcentrifuge and remove as much PBS 0.1%T as possible without disturbing the pellet.
5. Store dissected ovaries on ice for immediate use or freeze at −80 °C (*see* **Note 1**).
6. Add 300 µL of HQ low-salt buffer to the sample, grind the tissues on ice using motor-driven pestle for 1 min. Spin briefly to pellet tissues and crush again 1 min on ice.

7. Add 300 μL of HQ low-salt buffer and homogenize by pipetting gently. Centrifuge at 10,000 × *g* for 30 min at 4 °C to pellet insoluble debris (*see* **Note 2**).
8. Transfer supernatant (around 550 μL) in new 2-mL microcentrifuge tube, avoiding the pellet and the layer of lipids above aqueous phase.
9. Homogenize the clarified lysate by pipetting gently on ice and immediately inject it onto the HQ column.

3.2 Manual HQ Chromatography

Subsequent steps are performed at the bench, both materials and reagents have to be precooled at 4 °C, and chromatography buffers have to be stored on ice. A schematic representation of the manual chromatography procedure using syringes is shown in Fig. 1b (*see* **Note 3**).

1. Activate the charged matrix by injecting 5 mL of HQ regeneration buffer in the column with 10 mL syringe; discard the flow through (*see* **Note 4**).
2. Equilibrate the column by injecting 7 mL of HQ low-salt buffer with 10 mL syringe, discard the flowthrough (*see* **Note 4**).
3. Transfer the clarified lysate in 1 mL syringe, place the column adaptor on the syringe then push slowly to purge air and avoid any gas injection in the column. Inject the clarified lysate onto the column and discard the flow through. In this step, molecules are adsorbed on the matrix by electrostatic interaction.
4. Inject 900 μL of HQ low-salt buffer onto the column with 1-mL syringe and collect 900 μL of flow through in 2-mL microcentrifuge tube.
5. Add 100 μL of HQ high-salt buffer in the collected flow through and homogenize gently by pipetting. In order to minimize the volumes during chromatography, this mix is used as HQ mild-salt buffer for the first elution step.
6. Transfer the mix of the previous step in 1-mL syringe, inject it through the column and collect 1 mL of flow through in 2-mL microcentrifuge tube. Keep this elution fraction on ice. This is the first elution step called "Elution I" in Fig. 1b. In the elution steps, the electrostatic interaction is weakened allowing elution of the RISC complexes, whereas naked RNAs remain bound to the column.
7. Mix 900 μL of HQ low-salt buffer with 100 μL of HQ high-salt buffer in 2 mL microcentrifuge tube and homogenize by vigorous vortexing to obtain fresh HQ mild-salt buffer for the second elution step. Spin briefly and transfer the mix in 1 mL syringe then inject it through the column. Collect the 1 mL flow through in 2-mL microcentrifuge tube. This 1 mL fraction corresponds to "Elution II" in Fig. 1b.

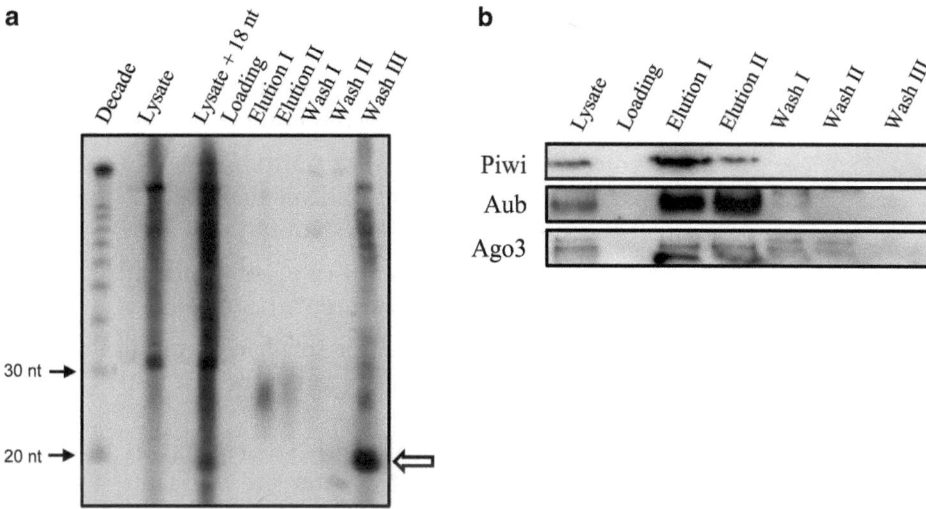

Fig. 2 Manual HQ chromatography enriches mild-salt elution I and II fractions in piRISCs. (**a**) ^{32}P radiolabeling RNA extracted from manual HQ purification fractions (*see* Fig. 1a) of Drosophila ovaries migrated on 15 % acrylamide gel. 18-nt long RNA oligomers have been added to native lysate prior to purification (Lysate + 18 nt lane) to mimic uncomplexed RNA. *White arrowhead* illustrates it recovery in high-salt washing fraction (Washing III lane). (**b**) Western blot performed on clarified lysate and on fractions retrieved during manual HQ purification procedure of Drosophila ovaries (*see* Fig. 1a) and revealed using antibodies against PIWI proteins

8. Mix elution I and elution II in 2-mL microcentrifuge tube, vortex vigorously, and spin briefly, then store on ice until RNA extraction. These elution fractions contain the majority of the PIWI proteins and their associated piRNAs (Fig. 2a, b).
9. Optional step: Recycling the column or/and studying the naked small RNA and proteins eluted at high-salt concentration (*see* **Note 5**) (Fig. 2a, b).

3.3 RNA Extraction from Pooled Eluted Fractions

RNAs are isolated from proteins using acidic phenol/chloroform extraction following by ethanol precipitation. Crude RNAs could be stored for month in −80 °C or immediately used. From this step, both materials and reagents have to be RNAse free (*see* **Note 6**).

1. Split the pooled elution fractions in four fractions of 500 μL in 1.5 mL microcentrifuge tubes.
2. Deproteinize the sample by adding 750 μL acidic phenol/chloroform (pH 4.5). Mix vigorously by shaking the tubes. Centrifuge at 10,000 × *g* for 10 min at 4 °C.
3. Transfer the aqueous (upper) phase to new 1.5-mL microcentrifuge tube without disrupting the lower phase.
4. Re-extract by adding 750 μL chloroform to avoid phenol contamination. Mix vigorously by shaking the tubes. Centrifuge at 10,000 × *g* for 5 min at 4 °C, then transfer the aqueous (upper) phase in new 2-mL microcentrifuge tube.

5. Add 55 μL 3 M sodium acetate (pH 5.5) and 1 μL of GlycoBlue, mix by tapping the tube and add 1.5 mL of cold absolute ethanol then homogenize by inverting the tubes 4–5 times (*see* **Note 7**). Incubate for minimum 4 h at −80 °C.
6. Centrifuge at 20,000 × *g* for 45 min at 4 °C to pellet RNAs. Discard the supernatant and incubate tubes on dry bath at 60 °C until pellet is dry (*see* **Note 8**).
7. Resuspend the first pellet in 15 μL of formamide and reuse this volume to resuspend the three other pellets.
8. RNAs can now be used for small RNA purification on polyacrylamide gel or stored several months at −80 °C.

3.4 Small RNA Polyacrylamide Gel Isolation

Purification of small RNAs for RNA-seq implies to concentrate small RNAs in DEPC-treated water and to clean the sample from residual salt prior to library construction. This step is performed by extracting 18–35-nt long small RNAs from polyacrylamide gel that allowed to obtain high-quality samples ready to send to the sequencing facility.

1. Set up a reaction mixture in 10 μL with 1 μL of each 10 mM RNA ladders, 1 μL of 10× kinase buffer, 1 μL of 32P ATP, and 5 μL of DEPC-treated water. Mix and spin the tube briefly then add 1 μL of T4 polynucleotide kinase, homogenize gently by tapping the tube, and spin briefly. Incubate for 1 h at 37 °C.
2. Homogenize G25 column resin by vortexing, then twist off the bottom enclosure and unlock cap by one quarter turn. Place the column on collecting tube, then centrifuge at 700 × *g* for 1 min at 4 °C. Equilibrate the column by loading 30 μL of DEPC-treated water and centrifuge at 700 × *g* for 1 min at 4 °C. Discard the collecting tube and replace it by new 1.5-mL microcentrifuge tube (*see* **Note 9**).
3. Add 10 μL of DEPC-treated water in the kinase reaction and mix gently by pipetting and apply the sample carefully in the center of the resin.
4. Centrifuge at 700 × *g* for 1 min at 4 °C, collect the eluted fraction and then add 6 μL of 4× RNA loading blue and mix gently by pipetting. This RNA ladder mix is ready to be loaded on acrylamide gel (*see* **Note 10**).
5. Prepare small glass plates (8 × 10 cm), spacers (1.5 mm thick) and comb for pouring gel in RNAse-free conditions (*see* **Note 11**).
6. In 15-mL Falcon tube, add 10 mL of acrylamide gel stock solution, 60 μL of 10 % APS and 5 μL of TEMED, mix well by inverting tube several times. Pour the gel immediately and add the appropriated comb. After 30 min at room temperature, the acrylamide is polymerized.

7. Assemble the gel in the electrophoresis unit then fill the upper and lower reservoirs with TBE 1× running buffer. Remove the comb and the excess polymerized gel that could interfere with loading. Rinse the urea from wells using syringe loaded with TBE 1×.
8. Pre-run the gel at 300 V for 30 min (*see* **Note 12**). Ensure that the upper reservoir does not leak.
9. Add 5 μL of RNA loading blue in each sample and homogenize by tapping tube, then spin briefly.
10. Denature secondary structures by heating sample and RNA ladders at 90 °C for 2 min and immediately chill on ice. Once tubes are cooled, spin briefly, homogenize by tapping tube and spin briefly again.
11. Just before loading samples on gel, rinse the urea from wells with TBE 1× loaded on syringe (*see* **Note 13**). Load sample carefully. Flank each sample to extract by RNA ladders mix (*see* **Note 14**).
12. Run the gel at 300 V for 25 min (*see* **Note 15**).
13. Disassemble the electrophoresis system and check the absence of radioactive signal in the running buffer or on the materials. Carefully remove the stackers and one of the glass plate, cover the gel with plastic sheet and then place fluorescent markers on the edge of the gel (*see* **Note 16**).
14. Expose the gel to autoradiography film in dark room and cut with clean scalpel blade the region of interest between 18 and 35 nucleotides based on the RNA ladders. Transfer the gel slice on clean glass plate and check the absence of radioactive signal.
15. Cut the gel in small fragments, and then transfer the pieces in 1.5-mL microcentrifuge tube.
16. Add 500 μL of 10 % 5 M NaCl in DEPC-treated water (at least three gel volumes), and crush manually the gel with a small pestle. Close the tube and wrap the top with parafilm then incubate on rotating wheel overnight at 4 °C.
17. Spin briefly to pellet the acrylamide debris and transfer the liquid phase onto Spin X 0.45 μm cartridge placed in new 1.5-mL microcentrifuge tube. Centrifuge at 3,800×*g* for 10 min at 4 °C and collect the flow through.
18. Add 1 μL of GlycoBlue, mix and add 500 μL of cold isopropanol. Homogenize by inverting tube. Incubate minimum 4 h to overnight at −80 °C.
19. Centrifuge at 20,000×*g* Äi0for 45 min at 4 °C, and carefully remove the supernatant. Dry rapidly the pellet in dry bath at 60 °C (*see* **Note 8**).
20. Solubilize small RNAs in 7 μL of 10 % RNasin in DEPC-treated water and store them at −80 °C.

3.5 HQ Small RNA Quality Control

This step is essential to control the presence of small RNAs after purification and the absence of sample degradation during the steps that follow chromatography. Radiolabeling followed by acrylamide gel electrophoresis is sensitive enough to detect a low amount of small RNAs. Quantification of the sample prior to perform the RNA library is also required (*see* **Note 17**).

1. Mix 1 μL of small RNA sample with 1 μL 10× dephosphorylation buffer, 7 μL DEPC-treated water and mix gently by tapping tube then spin briefly. Add 1 μL Calf Intestine alkaline phosphatase, mix gently and spin briefly. Incubate 1 h at 37 °C.
2. Add 10 μL of homogenized Phosphatase Removal Reagent (PRR), mix by tapping tube for 3 min at room temperature.
3. Spin briefly on microcentrifuge to pellet the PRR, and then carefully transfer the aqueous phase (15 μL) in new 1.5-mL microcentrifuge tube, avoiding PRR resin contamination that inhibits the kinase reaction (*see* **Note 18**).
4. Add 2 μL of 10× Kinase buffer, 1 μL of ^{32}P ATP, 1 μL DEPC-treated water and homogenize by tapping tube, then spin briefly. Add 1 μL T4 polynucleotide kinase, homogenize, and spin briefly. Incubate 1 h at 37 °C. At the end of the reaction add 10 μL of DEPC-treated water in tube and homogenize by pipetting.
5. Remove the unincorporated radio nucleotides using G25 column as described in **step 2** in Subheading 3.4 (*see* **Note 9**).
6. Add 6 μL of 4× RNA loading blue and homogenize by tapping tube then spin briefly.
7. Secondary structures are denatured by incubating 2 min at 90 °C, samples are immediately chilled on ice.
8. The quality of the radiolabeled RNAs can be simply evaluated by loading them into 15 % polyacrylamide gel under denaturing conditions (*see* **Note 11** and **steps 5–14** in Subheading 3.4).

4 Notes

1. Dissected ovaries can be stored at −80 °C for months. Freezing does not affect the manual HQ purification procedure.
2. During this step, the column can be regenerated and equilibrated for immediate use (*see* **Note 4**).
3. The column manufacturer provides adaptors for connecting the syringe to column. Carefully check the connection between syringe and column to avoid any air injection. During chromatography, the flow in the column has to be regular and do not exceed 1 mL/min.

4. For convenience, peristaltic pump can be used for regeneration and equilibration steps.
5. Wash the HQ column by injecting 1 mL of HQ regeneration buffer with 1 mL syringe. Repeat this step twice. Then wash the column with 1 mL of HQ high-salt buffer. Flowthroughs can be collected for further analysis as reported in Fig. 1b "Wash I, Wash II, and Wash III". The column could now be re-equilibrated to perform another purification (*see* **step 2** in Subheading 3.2), or stored for months at 4 °C after injection of 5 mL of column storage buffer.
6. Use aerosol barrier tips to prevent cross-contamination between samples and RNAse contamination.
7. Prepare the reaction mixture in the order indicated: sample, salt, GlycoBlue mix by tapping and then add alcohol and homogenize by inverting the tube several times.
8. Do not wash the pellet with 75 % ethanol because small RNAs could be lost in this step but remove carefully the maximum of supernatant.
9. Removal of unincorporated radiolabeled nucleotides allows to work with less radioactivity and to measure more precisely the signal of incorporated radiolabeled nucleotides.
10. Labeled RNA ladders in RNA loading blue can be stored several weeks at −20 °C.
11. This protocol is established for Mini PROTEAN 3 (BIORAD) and can be applied for similar 10 × 8 cm gel system. We recommend using small electrophoresis systems for this step in order to optimize the RNA elution from the gel. Larger gels that give better resolution are preferred for RNA sample quality control (*see* **step 8** in Subheading 3.5).
12. Pre-run is an optional step recommended to warm-up the gel and to obtain well-defined RNA bands for precise excision.
13. Wash to get rid of urea using TBE 1× is critical for a homogenous sample loading and migration.
14. The presence of RNA ladder lanes between samples minimizes the potential cross-contamination between samples that could occur during gel loading and gel excision.
15. This short migration time reduces the window of small RNA migration in the gel allowing a better RNA excision from the gel.
16. Three fluorescent marks on corners are enough to be precise when you will have to cut the gel. Perfect alignment between autoradiography and gel is a critical step for precise excision of the region of interest.

17. For small RNA quantification, do not use Nanodrop spectrophotometer due to the low sensibility range of this apparatus but use RNA Pico 6000 Arrays with Agilent technologies that provide both more precise quantification and the length profile of the purified material.
18. Due to the variable volume of liquid added to PRR resin, it is not always possible to recover 15 μL of aqueous phase but it is preferable to recover less volume without PRR resin, and complete to 15 μL with DEPC-treated water before the kinase reaction.

Acknowledgments

We thank members of the Chambeyron laboratory for helpful discussion, Anna Warnet for advices, A. Pelisson and N. Lamb for reading the manuscript, M.C. Siomi and J. Brennecke for antibodies. Grentzinger was the recipient of a fellowship from French Ministère de la Recherche et de l'enseignement supérieur (MRT). Work in the Chambeyron laboratory is supported by grants from l'Association pour la Recherche sur le Cancer (ARC JR/AD/DMV-09/2/5007), ANR Jeunes chercheurs (ESSOR-JCJC-1604 01), and CNRS.

References

1. Bingham PM, Kidwell MG, Rubin GM (1982) The molecular basis of P-M hybrid dysgenesis: the role of the P element, a P-strain-specific transposon family. Cell 29:995–1004
2. Bregliano JC, Picard G, Bucheton A, Pelisson A, Lavige JM, L'Heritier P (1980) Hybrid dysgenesis in Drosophila melanogaster. Science 207:606–611
3. Castro JP, Carareto CM (2004) Drosophila melanogaster P transposable elements: mechanisms of transposition and regulation. Genetica 121:107–118
4. Chambeyron S, Bucheton A (2005) I elements in Drosophila: in vivo retrotransposition and regulation. Cytogenet Genome Res 110:215–222
5. Brennecke J, Aravin AA, Stark A, Dus M, Kellis M, Sachidanandam R, Hannon GJ (2007) Discrete small RNA-generating loci as master regulators of transposon activity in Drosophila. Cell 128:1089–1103
6. Gunawardane LS, Saito K, Nishida KM, Miyoshi K, Kawamura Y, Nagami T, Siomi H, Siomi MC (2007) A slicer-mediated mechanism for repeat-associated siRNA 5′ end formation in Drosophila. Science 315:1587–1590
7. Brennecke J, Malone CD, Aravin AA, Sachidanandam R, Stark A, Hannon GJ (2008) An epigenetic role for maternally inherited piRNAs in transposon silencing. Science 322: 1387–1392
8. Chambeyron S, Popkova A, Payen-Groschene G, Brun C, Laouini D, Pelisson A, Bucheton A (2008) piRNA-mediated nuclear accumulation of retrotransposon transcripts in the Drosophila female germline. Proc Natl Acad Sci U S A 105:14964–14969
9. Lau NC, Seto AG, Kim J, Kuramochi-Miyagawa S, Nakano T, Bartel DP, Kingston RE (2006) Characterization of the piRNA complex from rat testes. Science 313:363–367
10. Lau NC, Robine N, Martin R, Chung WJ, Niki Y, Berezikov E, Lai EC (2009) Abundant primary piRNAs, endo-siRNAs, and microRNAs in a Drosophila ovary cell line. Genome Res 19:1776–1785
11. Grentzinger T, Armenise C, Brun C, Mugat B, Serrano V, Pelisson A, Chambeyron S (2012) piRNA-mediated transgenerational inheritance of an acquired trait. Genome Res 22: 1877–1888

Chapter 15

Isolation of Zebrafish Gonads for RNA Isolation

Hsin-Yi Huang and René F. Ketting

Abstract

Piwi proteins and piRNAs are abundant in the gonads of various animal species. Gonads from different developmental stages provide us information regarding the function of piRNAs and the PIWI pathway during germline development. Here we describe methods for gonad and germ cell preparation from different developmental stages of zebrafish. We also describe how to use these gonads to purify and characterize piRNAs.

Key words Zebrafish, piRNA, Germline development, Gonads, Ovary, Testis, RNA, FACS, Vasa

1 Introduction

Zebrafish (*Danio rerio*) is a tropical freshwater fish that has been used as a model organism for various research fields, such as regeneration, vertebrate development, gene functions, environmental sciences, and cancer models.

As a model organism, zebrafish possess several advantages (ZFIN website: http://zfin.org/). Primarily, zebrafish are easy to cultivate and take relatively little space to breed. Embryonic development is extremely fast, establishing most organs during the first 24–48 h, and occurs completely outside the mother. During the first days the embryos are transparent, allowing easy visualization of internal processes and live cell imaging combined with fluorescent labeling. This offers particular advantages and reduces experimental animal use compared to other model systems from which the embryos have to be dissected from the mother. Zebrafish take roughly 12 weeks to develop from fertilized eggs to sexually mature adults, although this time span can be influenced significantly by environmental conditions. Finally, they have a large clutch size. Generally, a pair of adult zebrafish can produce in excess of 100 eggs each time they mate, allowing parallel analysis of many siblings. The zebrafish genome is largely annotated and many molecular techniques are well-developed for visualization of

Mikiko C. Siomi (ed.), *PIWI-Interacting RNAs: Methods and Protocols*, Methods in Molecular Biology, vol. 1093,
DOI 10.1007/978-1-62703-694-8_15,

proteins and RNAs. There are techniques available for genetic manipulation, such as making transgenic animals using the Tol2 transposon system, as well as mutant animals from forward and reverse genetic screens (http://www.zfin.org). Taken together, these advantages make the zebrafish a valuable model system for developmental and genetics researchers.

In zebrafish, germ cell fate is determined by the inheritance of the germ plasm. Germ plasm is a maternally provided cytoplasm composed of RNA and RNA-binding proteins. It is first formed at the distal end of the cleavage furrows at the 2- and 4-cell stage embryos [1]. At the 32-cell stage, 4 cells have obtained the germ plasm aggregates and become promptive primordial germ cells (pPGCs) [2]. These cells undergo migration and proliferation and become primordial germ cells (PGCs). At 24 h post-fertilization (hpf), PGCs end up at the future gonad area, numbering between 25 and 50 cells [3].

Zebrafish are considered juvenile hermaphrodites. All fish first develop an immature ovary, called the juvenile ovary, during juvenile development. Later in development, the gonads differentiate into either ovaries or testes [4]. The timing of the gonadal developmental stages varies between different laboratory settings, but their sequence remains the same. In our laboratory, we observe that PGCs have differentiated into juvenile oocytes around 3 weeks post-fertilization (wkpf). At 5 wkpf, while most of the fish retain the juvenile ovaries, some possess gonads undergoing transformation into testes. The animals having higher numbers of juvenile oocytes are speculated to become future-females. Other animals are speculated to become males based on the following phenotypes: degeneration of the juvenile oocytes, higher numbers of gonadal somatic cells, and onset of spermatogenesis [5]. The transition from juvenile hermaphrodites to mature females or males can take up to 8 weeks. After 12 wkpf, most of the animals are sexually mature.

Zebrafish do not have sex chromosomes as many other species do. Sex determination of zebrafish depends on several factors, such as the environment and the population density. It has been proposed that the number of germ cells present in the early gonads also affects sex determination in zebrafish and mutants lacking germ cells develop into infertile males ([6–10]; Ketting, unpublished data). Zebrafish that are mutant for Piwi pathway components mostly display infertility due to germ cells loss. The developmental time at which germ cells are lost can vary, making analysis of germ cells at different developmental stages essential. In some cases it is even required to sort germ cells from the gonadal somatic tissue. Below we describe methods to isolate gonads and germ cells from zebrafish of different developmental stages and we provide protocols for RNA isolation and immune-precipitation for the isolation of piRNAs.

2 Materials (*see* Note 1)

2.1 Isolating Adult Zebrafish Gonads

1. White light dissection microscope with light source from the top.
2. 1.5-mL microcentrifuge tubes.
3. Insulated bucket with ice.
4. Trizol® LS.
5. #22 surgical blade or a pair of surgical scissors.
6. #11 surgical blade.
7. Two pairs of sharp-ended #5 surgical tweezers.
8. 10 cm petri dish.

2.2 Isolating 3–7 wkpf Gonads

1. Lysis buffer for genotyping: 10 mM Tris–HCl (pH 8.0), 50 mM KCl, 0.3 % Tween 20, 0.3 % NP40, 1 mM EDTA.
2. 20 mg/mL Proteinase K solution.
3. Light microscope with light source from the bottom.
4. Plate A: 96-well round bottom plates filled with 200 μl of cold PBS (without Mg^{2+} and Ca^{2+}) per well.
5. Plate B: 96-well PCR plates filled with 50 μl of lysis buffer per well.
6. Bucket with icy water.
7. #22 surgical blade.
8. Two pairs of sharp-ended #5 surgical tweezers.
9. 10 cm petri dish.

2.3 Isolating Gonads from 3-Week-Old Zebrafish

1. Light/Fluorescent microscope with the light source from the bottom.
2. Trizol® LS.
3. 30G needles.
4. 1 mL syringes.
5. One pair of sharp-ended # 5 tweezers.
6. 10 cm petri dish.

2.4 Germ Cells from 3-Week-Old Animals

1. Light/Fluorescent microscope with white light source from the bottom.
2. 28 °C water bath.
3. Desktop centrifuge for microcentrifuge tubes, set at 28 °C.
4. Vasa::EGFP expressing zebrafish.
5. MoFlo Fluorescence-Activated Cell Sorting machine (Dakocytomation).
6. 30G needles.

7. 1 mL syringes.
8. One pair of sharp-ended # 5 tweezers.
9. 10 cm petri dish.
10. P1000 tips.
11. 40 μm nylon cell strainer (BD Falcon).
12. Cold PBS without Mg^{2+} and Ca^{2+}.
13. TrypLE® Express (Invitrogen), Room temperature.
14. 100 % Fetal Calf Serum (FCS), Room temperature.

2.5 Total RNA Isolation from Gonads or Sorted Germ Cells

1. Desktop centrifuge for microcentrifuge tubes, precooled to 4 °C.
2. Cold Phenol/Chloroform solution.
3. RNase-free water.
4. 100 % RNase-free Ethanol.
5. Cold 75 % RNase-free Ethanol.
6. GlycoBlue® (Life Technologies).
7. FastRNA® Pro Green Kit (MP Biomedicals).
8. Plastic pestle fitting a 1.5-mL microcentrifuge tube.

2.6 Immuno-Precipitation

1. Cold IP lysis buffer: 25 mM Tris–HCl (pH 7.5), 150 mM NaCl, 1.5 mM $MgCl_2$, 1 % TritonX100, 1 mM DTT.
2. Cold IP wash buffer: 25 mM Tris–HCl (pH 7.5), 150 mM NaCl, 1.5 mM $MgCl_2$, 1 mM DTT.
3. cOmplete mini, EDTA-free (Roche).
4. rRNasin (Promega).
5. Antibodies against PIWI pathway components.
6. ProteinG Dynabeads (Novex).
7. Plastic pestle fits for 1.5-mL microcentrifuge tubes.
8. Bioruptor® (Diagenode).
9. Microcentrifuge tube rotator in a 4 °C cold room.
10. Desktop centrifuge for microcentrifuge tubes, precooled to 4 °C.
11. Magnetic rack for Dynabeads.

2.7 Small RNA Isolation from Gel

1. 15 % denaturing acrylamide gels (RNase-free).
2. RNase-free 0.5× TBE buffer: 45 mM Tris–Borate and 0.1 mM EDTA.
3. RNase-free 0.3 M NaCl.
4. RNase-free water.
5. 21G needle.
6. Spin-X cellulose Acetate filters.
7. SYBR® Gold Nucleic Acid Stain (Invitrogen).

3 Methods

3.1 Isolating Adult Gonads (See Note 2)

1. Prepare sample-collecting tubes on ice.
2. For one to three ovaries or six testes from adult zebrafish, prepare 500 μL of Trizol® LS in the microcentrifuge tubes.
3. Euthanize the animal to be dissected in ice–water.
4. Position the fish on its side on the petri dish.
5. Remove the part frontal from the end of the gills.
6. Remove the part distal from the anus.
7. Use one pair of tweezers to hold the remaining part of the fish trunk, ventral side facing up.
8. Cut open the ventral side using the #11surgical blade, cutting in the direction from frontal to distal of the fish trunk.
9. Gently peel the skin of the body with both pairs of tweezers.
10. In the trunks from female fish, many maturing oocytes can be observed covering the whole ventral side of the trunk. Use one pair of tweezers to hold open the skin while inserting the other pair of tweezers gently in between the frontal ovary and the backbone of the trunk. Gently pull out the ovary in the direction from frontal to distal of the fish trunk. During the process of pulling out the ovary, you will observe the cluster of guts and liver tissues. Separate the guts gently with the tweezers. Do not damage the guts to prevent leakage of RNase from the digestive tissues. Try not to puncture the thin membrane around the ovary.
11. In the trunks from male fish, initially the cluster of guts and livers will be observed. Remove the cluster gently and discard. In the middle of the remaining trunk, there is the long strip of swim bladder. At each caudal side of the swim bladder, there is one lobe of testis. These two lobes of testes are connected to each other at their distal ends. Use the tweezers to hold this connecting part and pull off gently the testes.
12. The swim bladder may come off with the gonads. After taking the gonads out of the fish trunk, leave the gonads on a clean spot of the petri dish. Gently pull off the swimming bladder before placing the gonads in the sample tubes containing cold Trizol® LS.

3.2 To Genotype 3–7 wkpf Animals Prior to Isolation of the Gonads

1. Add Proteinase K solution (20 mg/mL) to lysis buffer in 1:100 dilution right before adding the lysis buffer to Plate B.
2. Keep all 96-well plates on ice.
3. Euthanize a fish in icy water.
4. Cut off the part of the fish distal from the anus and leave this part in a well of plate B.

5. Place the remainder of the fish body in the corresponding well in plate A.
6. After all the wells are filled or all samples processed, cover plate A with a lid or saran wrap. Leave it in a 4 °C fridge. Samples in plate A are stable for 4 h.
7. Prepare plate B for a lysis program in a PCR machine.
8. Lysis program: 60 min at 60 °C and followed by 15 min at 95 °C. Then store at 4 °C.
9. After the lysis program, add 2–3 times volume of water into the lysate, depending on the primer pairs for the following SNP genotyping.
10. Follow any genotyping protocol that allows determination of genotype within hours. We often use the KASPar protocol from KBioscience®.
11. After the genotype of each individual fish has been determined, one can proceed to the gonad isolation protocol.

3.3 Isolating 5–7 wkpf Gonads

In general, the procedures are the same as for adult gonads. The only thing to be noted is that at the age of 5–7 weeks of development differences between male and female gonads are poorly discernible under the dissection light microscope. The cluster of guts and liver tissues needs to be removed first prior to collecting the gonads from the dissected fish trunks.

3.4 Isolating Gonads from 3-Week-Old Zebrafish (See Notes 3 and 4)

1. Precool the microcentrifuge tubes on ice.
2. For every 80–100 3-week-old gonads, prepare 500 μL of Trizol® LS per microcentrifuge tube.
3. Assemble the needles and syringes. The needles will be used as mini knifes while the syringes are the holders of knifes.
4. Position the fish on its side; the fish will look like the schematic shown in Fig. 2. The swim bladder is located in the middle of the trunk as a long oval-shaped object stretching from the frontal to distal side of the trunk. The dorsal side of the swim bladder is mostly muscle and is less transparent. The ventral side of the swim bladder is adjacent first to the gonads and then the cluster of guts. The guts will appear darker or colored under the microscope due to remaining food.
5. Using one pair of the tweezers to stabilize the fish while using a needle knife to cut off the guts and the tissue at the ventral side of the guts. The movement of the needle knife is from frontal to distal. Be aware that the distal end of the guts is often tightly connected to the distal end of the gonads. Gently separate the guts and the gonads with the needle knife.
6. Discard the guts. And then cut off the swim bladder together with the gonads using the needle knife.

7. Work with two needle knifes together to further separate the skin and swim bladder from the gonads. Transfer the gonads into Trizol® LS as soon as possible.

3.5 Prepare Samples for Fluorescence-Activated Cell Sorting (FACS)

1. The protocol for gonad isolation is as described in Subheading 2.4. However, gonads with Vasa::EGFP transgene are collected in PBS instead of Trizol®.
2. After the gonads are collected, spin down the tissue at 4 × *g* for 1 min.
3. Remove the PBS and replace it with 500 μL of TrypLE® Express and keep the samples in a 28 °C water bath.
4. Gently resuspend the tissues using p1000 tips every 10 min.
5. Check the dissociation once in a while under the microscope.
6. For 80 gonads, it takes about 30 min to dissociate. It also depends on how much somatic tissue remained around the gonads.
7. Add 120 μL of 100 % FCS and gently mix the solution to stop the trypsin reaction.
8. Spin down the cells at 4 × *g* for 1 min.
9. Remove the supernatant and resuspend the cells in cold PBS.
10. Pass the cells through a 40 μm nylon cell strainer and keep the samples on ice prior to FACS.
11. High GFP cells are sorted directly into Trizol® LS.
12. The sample is then ready for RNA isolation.

3.6 Primordial Germ Cells (PGCs) from 1 to 7-Day-Old Embryos (See Note 5)

A few hundred embryos are required for sub-microgram amounts of RNA.

1. Embryos with labeled PGCs are required. We often use the vasa::EGFP transgene [11].
2. After euthanizing the embryos in icy water, collect the embryos in 1.5-mL microcentrifuge tubes.
3. Try to remove as much liquid as possible. If necessary, spin the embryos briefly.
4. The FACS sample preparation is similar to that described for 3-week-old gonads.
5. For every 200 embryos, use 500 μL of TrypLE® Express to dissociate the cells.
6. At 28 °C, it takes about 60 min to dissociate the cells from the tissue.
7. Gently resuspend the sample every 10–15 min. It is crucial to check once in a while under fluorescent microscope to ensure efficient dissociation without stressing the PGCs.

3.7 Total RNA Isolation from Adult Gonads or Sorted Germ Cells

Microgram amounts of RNA can be obtained from just a few individuals.

1. If the samples are frozen, let them thaw completely before proceeding further.
2. Grind the gonads in the Trizol LS® with the plastic pestle until there are no obvious chunks of tissues.
3. Follow the RNA extraction protocol according to the Trizol LS® manual.
4. After Trizol RNA extraction, dissolve the RNA pellet in 50 μL of RNase-free water and leave the sample on ice as much as possible.
5. Add additional 200 μL of RNase-free water and 300 μL of cold Phenol–Chloroform solution. Mix the solution by vortexing for a few seconds.
6. Centrifuge the samples at the maximum speed (20 × *g*) for 30 min.
7. Collect the upper aqueous phase and transfer it to a clean microcentrifuge tube. Discard the old sample tubes.
8. Add two sample volumes of 100 % Ethanol and 1 μL of GlycoBlue into a new microcentrifuge tube. Vortex for a few seconds to mix.
9. Keep the sample in a −80 °C freezer for 2 h for precipitation. After the precipitation step, centrifuge the samples at the maximum speed for 30 min.
10. Remove the supernatant and let the pellet air dry a bit before dissolving the RNA pellet in the desired amount of RNase-free water.

3.8 Total RNA Isolation from 3 to 7 wkpf Gonads and Sorted Germ Cells

To obtain microgram amounts up to 100 individuals are required. Skip the first two steps in Subheading 3.7 to reduce the loss of samples since fish tissues stick relatively easy to plastics. Instead of the Trizol LS® manual, use FastRNA® Pro Green Kit and the manufacturer's protocol followed by **steps 4–10** in Subheading 3.7.

3.9 Immuno-Precipitation

1. Freshly add DTT, cOmplete mini tablet (1 tablet per 10 mL buffer) and rRNasin (1 μl per 1 mL buffer) to the IP lysis buffer and keep on ice.
2. The procedure to isolate gonads for immune-precipitation is identical as for RNA isolation. The only exception is to replace the Trizol® LS solution with cold IP lysis buffer.
3. For every 1/2 to 1 adult ovary or 3 to 6 adult testes, use 500 μL of cold IP lysis buffer.

4. Grind the gonads with plastic pestle until no trunks of tissue are visible.
5. Set the Bioruptor® at High with 1 min ON and 30 s OFF.
6. Sonicate the samples in the ice/water bath of the Bioruptor for a total of 3 min.
7. Centrifuge the lysate at 20 × *g* for 10 min, 4 °C.
8. The lysate should be cloudy. Testes lysate often has a whitish layer on the lysate surface. This is normal and does not affect further steps.
9. Transfer the supernatant into a clean microcentrifuge tube.
10. Add desired amount of antibody and incubate for 2 h on the rotator in the cold room.
11. In the meanwhile, prepare ProteinG Dynabeads.
12. Wash the beads twice with cold IP lysis buffer.
13. After 2 h of incubation, add 30 μL beads per 500 μL lysate.
14. Let it incubate for additional 1 h on the rotator in the cold room.
15. Remove the supernatant using the magnetic rack for Dynabeads.
16. Wash three times briefly with the cold IP wash buffer.
17. Wash another three times with 5 min incubation on ice each. Do not let the beads fall dry during the washing steps.
18. Remove the wash buffer and proceed to protein or RNA isolation.

3.10 RNA Isolation After Immune-Precipitation

1. To isolate total RNA after immuno-precipitation, resuspend the beads in 300 μL of Trizol® LS and add 50 μL of RNase-free water.
2. Vortex to mix and then follow the RNA isolation protocol as mentioned above.

3.11 Small RNA Isolation from Gel

1. Heat sample with loading buffer at 65 °C for 5 min and snap cool on ice. Load approximately 10 μg of total RNA on denaturing acrylamide gel.
2. Load marker RNAs. We use 19, 24, and 34 nucleotide synthetic RNAs.
3. Run at 100 V until the bromophenol blue reaches the bottom of the gel.
4. Stain gel with SybrGold.
5. Cut out the desired size range from the gel and transfer to 0.5 mL tubes that has been punctured 3–4 times with the 21G needle.

6. Place this tube into a 2 mL tube and spin the gel slice through the whole at full speed for 2 min.
7. Add 500 μL 0.3 M NaCl to the disrupted gel slice and elute RNA by rocking overnight.
8. Transfer to Spin-X cellulose Acetate filter and spin at full speed for 2 min. Rinse gel fragments once more with 100 μL 0.3 M NaCl and combine eluates.
9. Precipitate RNA by adding one equal volume of isopropanol and glycoblue (to facilitate precipitation and for detection of pellet). Place at −20 °C for 1 h.
10. Spin 18 × *g* for 25 min in a cooled centrifuge (4 °C).
11. Carefully remove supernatant and wash pellet with 750 μL 75 % ethanol (room temperature). Air dry pellet and dissolve in 5–6 μL water.

4 Notes

1. All the steps should be kept on ice or done in the 4 °C cold room as much as possible to prevent protein and/or RNA degradation.
2. Isolated gonads in Trizol® LS can be kept in a −80 °C freezer for up to 3 months prior to proceeding to isolate RNA.
3. Gonads from 3-week-old animals are very tiny. It is therefore much easier to work with animals carrying a transgene that fluorescently label the germ cells, such as the vasa::EGFP transgene [11], which is highly expressed in the gonads throughout development.
4. At the age of 3 weeks, zebrafish are still relative transparent. Therefore, it is easier to distinguish the body parts if using a light source from the bottom of the microscope while isolating gonads. For further visual guidance *see* Fig. 1, showing a

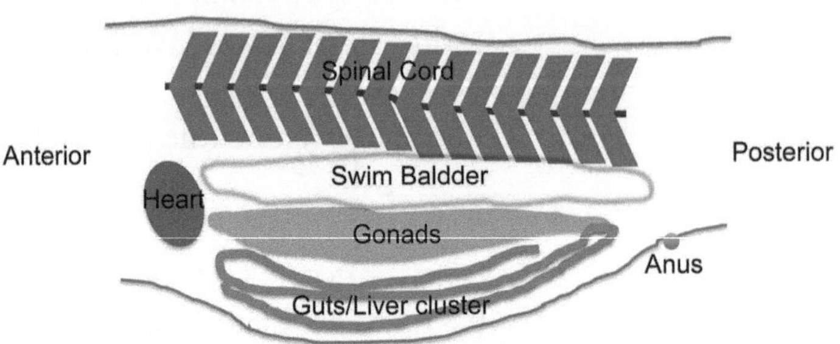

Fig. 1 Schematic side view of the body trunk of a 3-wkpf zebrafish. Drawing is not to scale

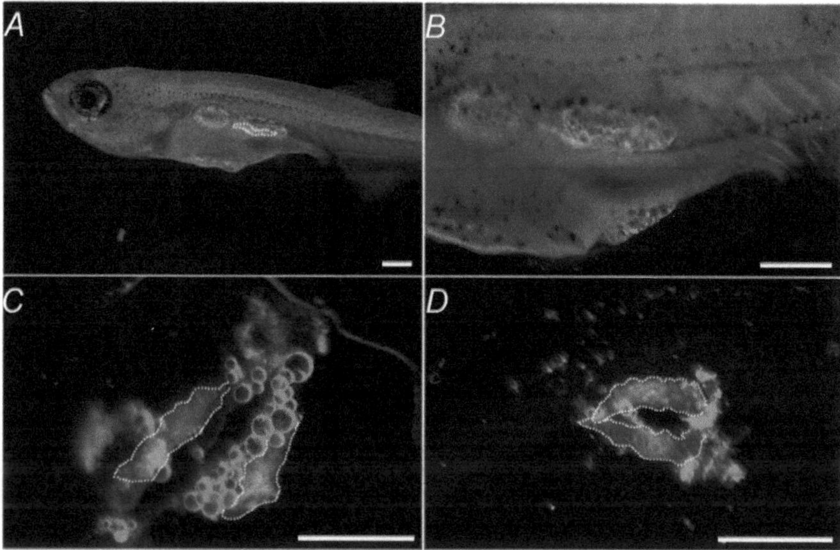

Fig. 2 Microscopic pictures of a 3 wkpf zebrafish and isolated gonads. (**a**) Overview of the whole fish from its side. (**b**) As (**a**) but zoomed in. In this case the skin was also removed. (**c**) Isolated gonads with some fat cells associated. (**d**) Isolated gonads after fat cells were cleaned away. Gonads (with green fluorescence signal) are indicated with *dashed line*. Scale bars are 50 μm

schematic drawing of the animal and Fig. 2, depicting representative microscopic pictures before and after isolation of the gonads from a 3-week-old fish.

5. This procedure works with earlier stage embryos as well, although usage of a different transgenic strain may give reduced fluorescent background compared to the vasa:EGFP line. A line suitable for PGC isolation from early embryos is kop-EGFP-F-nos1-3′UTR [12].

Acknowledgments

We thank Stefan Redl for help in making figures. This work was supported by VIDI (864.05.010), ECHO (700.57.006), and VICI (724.011.001) grants from the Netherlands Organization for Scientific Research.

References

1. Blaser H, Eisenbeiss S, Neumann M, Reichman-Fried M, Thisse B, Thisse C, Raz E (2005) Transition from non-motile behavior to directed migration during early PGC development in zebrafish. J Cell Sci 118:4027–4038
2. Braat AK, Speksnijder JE, Zivkovic D (1999) Germ line development in fishes. Int J Dev Biol 43:745–760
3. Houwing S, Berezikov E, Ketting RF (2008) Zili is required for germ cell differentiation and meiosis in zebrafish. EMBO J 27: 2702–2711
4. Houwing S, Kamminga LM, Berezikov E, Cronembold D, Girard A, van den Elst H, Filippov DV, Blaser H, Raz E, Moens CB et al (2007) A role for Piwi and piRNAs in germ

cell maintenance and transposon silencing in Zebrafish. Cell 129:69–82

5. Huang HY, Houwing S, Kaaij LJ, Meppelink A, Redl S, Gauci S, Vos H, Draper BW, Moens CB, Burgering BM et al (2011) Tdrd1 acts as a molecular scaffold for Piwi proteins and piRNA targets in zebrafish. EMBO J 30:3298–3308
6. Krovel AV, Olsen LC (2002) Expression of a vas::EGFP transgene in primordial germ cells of the zebrafish. Mech Dev 116:141–150
7. Raz E (2003) Primordial germ-cell development: the zebrafish perspective. Nat Rev Genet 4:690–700
8. Siegfried KR (2010) In search of determinants: gene expression during gonadal sex differentiation. J Fish Biol 76:1879–1902
9. Siegfried KR, Nusslein-Volhard C (2008) Germ line control of female sex determination in zebrafish. Dev Biol 324:277–287
10. Slanchev K, Stebler J, de la Cueva-Mendez G, Raz E (2005) Development without germ cells: the role of the germ line in zebrafish sex differentiation. Proc Natl Acad Sci USA 102: 4074–4079
11. Takahashi H (1974) Juvenile hermaphroditism in the Zebrafish, Brachyodanio rerio. Bull Fac Fish Hokkaido Univ 28:57–65
12. Yoon C, Kawakami K, Hopkins N (1997) Zebrafish vasa homologue RNA is localized to the cleavage planes of 2- and 4-cell-stage embryos and is expressed in the primordial germ cells. Development 124:3157–3165

Chapter 16

Small RNA Library Construction for High-Throughput Sequencing

Jon McGinn and Benjamin Czech

Abstract

Since their discovery about 20 years ago, small RNAs have been shown to play a critical role in a myriad of biological processes. The greater availability of high-throughput sequencing has been invaluable to furthering our understanding of small RNAs as regulatory molecules. In particular, these sequencing technologies have been crucial in understanding the role of small RNAs in reproductive tissues, where millions of individual sequences are generated. In this context, high-throughput sequencing provides the requisite level of resolution that other procedures, like northern blotting, would not be able to achieve. Here, we describe a protocol for the preparation of small RNA libraries for sequencing using the Solexa/Illumina technology.

Key words Endo-siRNA, miRNAs, piRNAs, Small RNA libraries, Cloning, High-throughput sequencing

1 Introduction

Eukaryotes utilize small RNAs to control a variety of biological processes ranging from gene regulation to guarding the genome from viruses and transposons. Nearly 20 years ago, evidence of a regulatory small RNA, *lin-4*, was uncovered [1, 2]. In the early 2000s, several novel small RNAs were identified [3–5], with multiple papers published describing new types of small RNAs since then. Differences in size, biogenesis, and Argonaute protein binding partners divide these small RNAs into three major classes: microRNAs, small interfering RNAs (siRNAs), and Piwi-interacting RNAs (piRNAs).

High-throughput sequencing has proven to be a powerful tool in profiling the entire repertoire of small RNAs in a tissue- or cell-specific context. This technology has been especially useful in reproductive tissues where small RNA populations are extremely complex and has significantly contributed to our understanding of piRNAs as a distinct class of small RNA [6–13]. The protocol described here is for the preparation of small RNA libraries for

Mikiko C. Siomi (ed.), *PIWI-Interacting RNAs: Methods and Protocols*, Methods in Molecular Biology, vol. 1093,
DOI 10.1007/978-1-62703-694-8_16,

sequencing using Solexa/Illumina technology (with 5′ monophosphate and 3′ hydroxyl groups), with modifications to previously reported protocols [4, 14, 15]. While other protocols and kits are available, this protocol describes a cost-effective and easily adaptable method for cloning small RNAs. The protocol can be modified for use on other sequencing platforms (e.g., SOLiD from ABI or 454 from Roche) by substituting the appropriate oligonucleotides and can be used for multiplexed sequencing by adding barcoding primers. It can also be adjusted to enrich for other classes of small RNA by including steps that create the proper termini.

2 Materials

RNase contamination could lead to failure of the protocol, therefore it is highly recommended to designate a clean working area and separate reagents to be used only when working with RNA. Also, be sure to follow the proper practices for work with radioactive materials (i.e., wearing the appropriate safety gear, waste disposal, etc.). General equipment, like a vortexer, pipets, tips, or centrifuges, is not listed. Materials used in several steps are only listed once. Materials and reagents can be stored at room temperature unless otherwise specified. The sequences of oligonucleotides are listed in Table 1.

2.1 Preparation of Ladder and Size Selection Markers

1. RNase-free 1.7-mL siliconized low-retention microcentrifuge tubes.
2. Nuclease-free water.

Table 1
Oligonucleotides for library preparation

Name	Sequence (5′ → 3′)
oLig3	/5rApp/CTGTAGGCACCATCAAT/3ddC/
oLig5	rArCrArCrUrCrUrUrUrCrCrCrUrArCrArCrGrArCrGrCrUrCrUrUrCrCrGrArUrC
oPCR3	CAAGCAGAAGACGGCATACGATTGATGGTGCCTACAG
oPCR5	AATGATACGGCGACCACCGAACACTCTTTCCCTACACGACG
o19M	rCrGrUrArCrGrGrUrUrUrArArArCrUrUrCrGrA
o24M	rCrGrUrArCrGrGrUrUrUrArArArCrUrUrCrGrArArArUrGrU
o28M	rCrGrUrArCrGrArUrCrCrGrUrUrUrArArArCrCrArUrUrGrUrUrCrArA

Nucleotides are DNA unless preceded by an "r", which indicates RNA. "*oLig3*" is a custom oligonucleotide that is pre-adenylated at the 5′ terminus and blocked at the 3′ terminus by a di-deoxy base (available through Integrated DNA Technology (IDT) under "miRNA cloning linker 1"). All oligonucleotides should be HPLC purified and stored at −20 °C. *Underlined* sequences correspond to PmeI restriction sites

3. Decade Markers with 10× cleavage reagent (Ambion). Store at −20 °C.
4. [γ-^{32}P]-ATP (6,000 Ci/mmol).
5. T4 Polynucleotide Kinase (PNK) (10,000 U/mL) with 10× T4 PNK Reaction buffer. Store at −20 °C.
6. RNA oligonucleotides *o19M* and *o28M* (each at 10 μM) (*see* **Note 1**).
7. Illustra MicroSpin G-25 Columns (GE Healthcare Life Sciences).
8. 0.1 mM ethylenediaminetetraacetic acid (0.1 mM EDTA).

2.2 Size Selection of Small RNAs and Gel Purification

1. SE400 Sturdier Vertical Electrophoresis Unit, 18 × 16 cm with 1-mm and 0.75-mm thick spacers, and 1-mm and 0.75-mm thick 15-well combs (Hoefer). Other polyacrylamide gel electrophoresis systems can be used.
2. UreaGel system (National Diagnostics). Other urea-containing, TBE-based gels (19:1 acrylamide:bisacrylamide) can be used.
3. 50 mL Falcon tubes.
4. 10 % ammonium persulfate (10 % APS). Store at 4 °C for up to 6 weeks.
5. *N*,*N*,*N*′,*N*′-tetramethylethylenediamine (TEMED).
6. 1× TBE (Tris/Borate/EDTA) electrophoresis buffer.
7. Syringe with a 22-gauge 1-in. needle.
8. 2× gel loading buffer (formamide-based with xylene cyanol and bromophenol blue).
9. RNase-free 200 μL gel loading tips.
10. Whatman 3MM Chr paper (GE Healthcare).
11. Adhesive tape.
12. Plastic wrap (Saran).
13. Phosphor screen and phosphorimager to visualize radiolabeled nucleic acids.
14. Printer.
15. Razor blades or scalpels.
16. 400 mM sodium chloride (400 mM NaCl).

2.3 Cleanup of Small RNA Samples

1. Micropore 0.22 μm filter microcentrifuge tubes (Ultrafree-MC; Millipore).
2. GlycoBlue (15 mg/mL dyed glycogen; Ambion). Store at −20 °C.
3. 100 % ethanol.
4. 80 % ethanol.

2.4 Ligation of 3′ Linker and Gel Purification of Ligated Product

1. *oLig3* oligonucleotide (100 μM).
2. Dimethylsulfoxide (DMSO), molecular biology grade.
3. Recombinant RNasin Ribonuclease Inhibitor (40 U/μL; Promega). Store at −20 °C.
4. T4 RNA Ligase 2, truncated K227Q (200,000 U/mL) with 10× T4 RNA Ligase Reaction Buffer, ATP-free (New England BioLabs). Store at −20 °C. Other truncated versions of T4 RNA ligase (from other vendors or homemade) can be used.

2.5 Ligation of 5′ Linker and Gel Purification of Ligated Product

1. *oLig5* oligonucleotide (100 μM).
2. T4 RNA Ligase (5 U/μL) with 10× T4 RNA Ligase Buffer (Ambion). Store at −20 °C.

2.6 Reverse Transcription

1. PCR thermal cycler.
2. RNase-free 0.2 mL PCR tubes.
3. *oPCR3* primer (100 μM).
4. dNTP mix (10 mM each). Store at −20 °C.
5. SuperScript III Reverse Transcriptase (200 U/μL) with 0.1 M DTT and 5× First-Strand Buffer (Invitrogen). Store at −20 °C.

2.7 PCR Amplification of cDNA

1. *oPCR5* primer (100 μM).
2. *oPCR3* primer (100 μM).
3. KOD Hot Start DNA Polymerase (1 U/μL) with 10× PCR Buffer, dNTP mix (2 mM each), and 25 mM $MgSO_4$ (EMD/Novagen). Store at −20 °C.
4. 3 M sodium acetate, pH 4.8 (3 M NaOAc, pH 4.8).

2.8 PmeI Digest and Gel Purification of Amplified cDNA

1. Restriction enzyme PmeI (10,000 U/mL) with 100× BSA and 10× NEBuffer 4 (New England BioLabs). Store at −20 °C.
2. Agarose for DNA recovery.
3. 1× TAE (Tris/Acetate/EDTA) electrophoresis buffer.
4. Ethidium bromide (EtBr) solution (10 mg/mL). Store protected from light.
5. GeneRuler 50 bp DNA Ladder supplied with 6× DNA loading dye (Fermentas). Other 6× loading dyes can be used. Store at 4 °C.
6. Long-wave UV transilluminator.
7. Wizard SV Gel and PCR Clean-up System (Promega). Gel purification kits from other vendors can be used.
8. NanoDrop or other spectrophotometer.

3 Methods

This protocol is used to clone small RNAs with 5′ monophosphate and 3′ hydroxyl groups, similar to other reported protocols [4, 14, 15]. Using a truncated version of T4 RNA ligase 2, small RNAs with 2′-*O*-methyl groups (like *Drosophila* endo-siRNAs and piRNAs) can be cloned with efficiencies similar to non-modified RNAs (like miRNAs) [16]. To clone small RNAs with other modifications (e.g., 5′ triphosphate or 3′ phosphate), the starting material can be treated with appropriate enzymes to convert their termini to 5′ monophosphate and 3′ hydroxyl groups.

In this protocol (*see* Fig. 1a), small RNAs of the desired size range are first isolated. Then, the 3′ adapter is ligated to the RNA and the ligation products are purified. The 5′ adapter is ligated next, and the RNA is gel purified again. The RNA is then converted to cDNA by reverse transcription, which then serves as a template for PCR amplification. The size selection markers are removed by PmeI restriction digest (*see* **Note 1**), and the library preparation is completed by agarose gel extraction. It is recommended to clone up to eight libraries per gel using a 15-well comb and running samples in every other lane to avoid cross-contamination. If more than eight libraries must be prepared in parallel, it is recommended to use multiple gels to safely accommodate all samples separated by a free or ladder lane. While this protocol is intended for constructing individual libraries, it can easily be adapted for multiplexed sequencing (*see* **Note 2**). For the best results, take the necessary precautions to minimize the risk of RNase contamination. Carry out all incubations at room temperature unless otherwise specified.

3.1 Preparation of Ladder and Size Selection Markers

1. Prepare the RNA ladder (Decade Markers) according to the manufacturer's instructions. Set up the following reaction: 6 μL of nuclease-free water, 1 μL of 10× kinase reaction buffer, 1 μL [γ-^{32}P]-ATP (6,000 Ci/mmol), 1 μL of Decade Marker RNA (100 ng), and 1 μL of T4 polynucleotide kinase in a nuclease-free microcentrifuge tube. Briefly mix and incubate for 1 h at 37 °C. Then add 8 μL of nuclease-free water and 2 μL of 10× cleavage reagent. Incubate for 5 min at room temperature. Then dilute by adding 30 μL of nuclease-free water and 50 μL of 2× gel loading buffer.
2. To prepare the RNA size selection markers (*see* **Note 1**), set up the following labeling reaction for the *o19M* and *o28M* oligonucleotides in separate nuclease-free microcentrifuge tubes: 14 μL of nuclease-free water, 0.5 μL of 10 μM RNA oligonucleotide (*o19M* or *o28M* from Table 1), 2 μL of 10× PNK buffer, 2.5 μL of [γ-^{32}P]-ATP (6,000 Ci/mmol), and 1 μL of PNK.

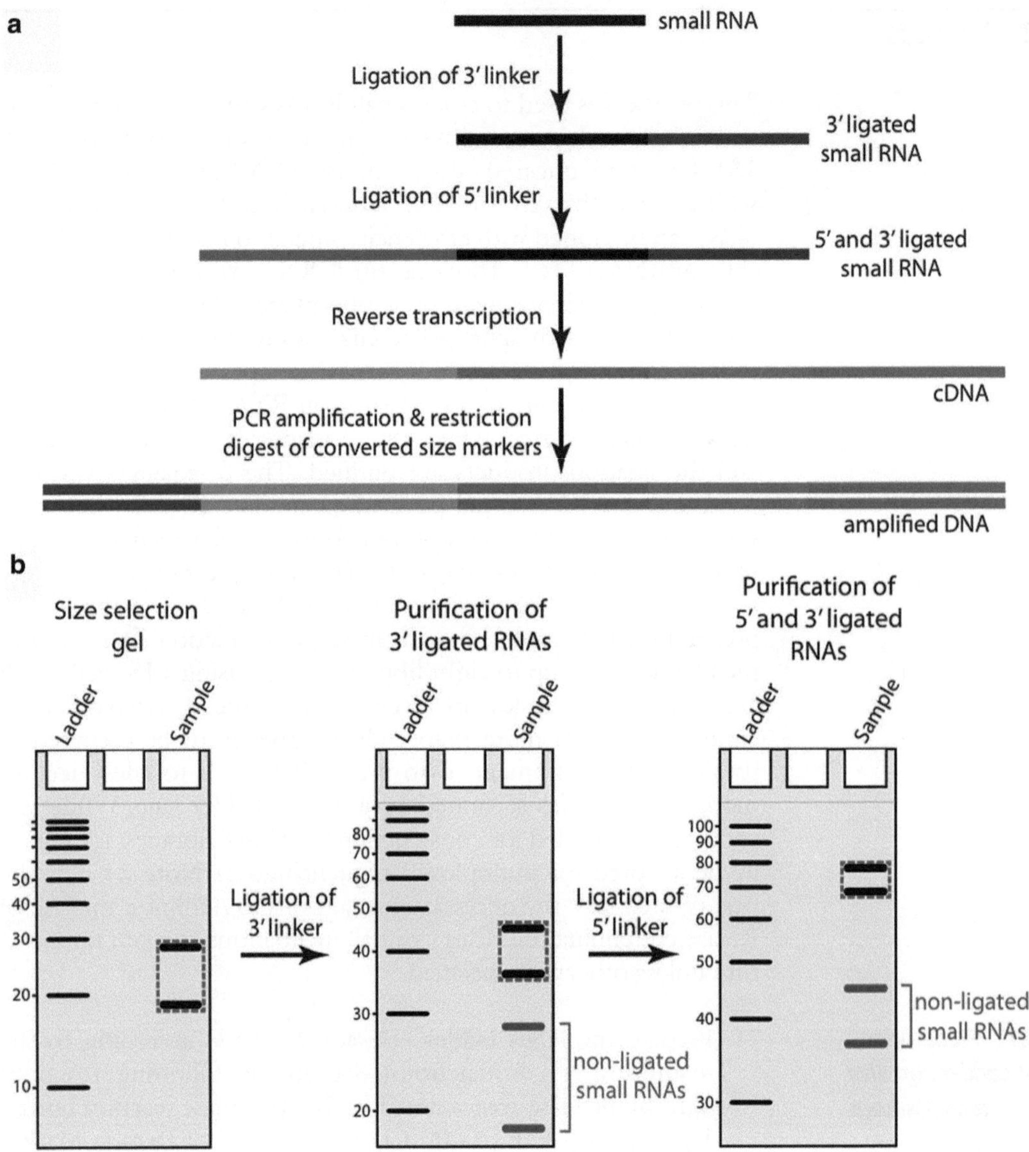

Fig. 1 Schematic overview of cloning strategy. (**a**) Small RNAs are size selected using cloning size selection markers. Extracted small RNAs are ligated with the 3′ linker and the appropriately sized fragments are gel purified. Following 5′ ligation, the appropriately sized fragments are again purified using denaturing PAGE. Properly ligated small RNAs are reverse transcribed and PCR amplified. The cloning size selection markers are removed by PmeI restriction digest followed by agarose gel purification to yield the sequencing-ready small RNA library. (**b**) Scheme of PAGE for size selection (*left*), selection of correct fragments following the 3′ ligation (*middle*), and selection of correct fragments following the 5′ ligation (*right*) are shown. A fraction of small RNAs does not ligate as indicated

3. Mix and incubate the reactions for 1 h at 37 °C, then heat-inactivate the enzyme by incubation at 65 °C for 20 min.
4. Remove unincorporated radiolabeled nucleotides using illustra MicroSpin G-25 columns according to the manufacturer's

instructions. In brief, resuspend beads by vortexing, then centrifuge the G-25 columns in the provided collection tubes at 735 ×*g* for 1 min. Discard the collection tubes and transfer the G-25 columns to clean microcentrifuge tubes. Add 30 μL of 0.1 mM EDTA to each heat-inactivated sample. Apply mixture to pre-processed G-25 column and centrifuge at 735 ×*g* for 2 min. The eluate contains radiolabeled markers.

5. Add one volume of 2× gel loading buffer to each labeled size marker.
6. *Optional: to ensure the quality of the ladder and size selection markers, run them on a 10% denaturing polyacrylamide gel using the UreaGel system* (*see* **step 2** in Subheading 3.4 for instructions).

3.2 Size Selection of Small RNAs and Gel Purification

1. Start with 10 μg of total RNA for each sample (*see* **Note 3**).
2. Mix each RNA sample with one volume of 2× gel loading buffer. Add 0.25 μL of each radiolabeled size marker (typically ~10,000 counts per minute of each size marker) from **step 5** in Subheading 3.1 to each sample. Dilute 2 μL of radioactive ladder in 40 μL 1× gel loading buffer.
3. Prepare a 1-mm thick, 12 % denaturing polyacrylamide gel using the UreaGel system according to the manufacturer's instructions. In brief, clean glass plates, spacers, and comb with water and ethanol, then assemble accordingly. For 50 mL of gel, mix 24 mL of concentrate, 21 mL of diluent, and 5 mL of 10× buffer (all included in the UreaGel system) in a Falcon tube. To start the polymerization reaction, add 400 μL of 10 % APS and 20 μL of TEMED, mix and immediately cast the gel. Insert the comb and allow the gel to polymerize for approximately 30 min. Once polymerized, carefully remove the comb and assemble the gel apparatus for running using 1× TBE running buffer. Make sure to rinse the wells using a syringe to remove residual urea. Pre-run the gel at 400 V for ~30 min.
4. Denature the samples and ladder (from **step 2** in Subheading 3.2) by heating for 2 min at 95 °C, and then immediately transfer to ice for 2 min. Briefly centrifuge to collect the sample at the bottom of the tube.
5. Following the pre-run, rinse the wells again using a syringe. Immediately load the samples and ladder using gel loading tips. Run the gel at 400 V until the bromophenol blue dye reaches bottom of the gel (~2 h).
6. Disassemble the gel apparatus, while keeping the gel on one glass plate.
7. Pipet 0.25 μL of radioactive ladder onto four small pieces of Whatman paper. Use adhesive tape to attach the radioactive paper pieces onto the edges of the glass plate. Carefully cover the gel and glass plate with plastic wrap.

8. Expose the gel to a phosphor screen for approximately 5 min (time depends on the amount of radioactivity used), and then scan the image with a phosphorimager and print the gel image (select actual size or 1:1 scale).
9. Place the printout under the glass plate and use the radioactive Whatman paper pieces to properly align the gel and printout.
10. For each sample, precisely excise the desired gel fragment (19–28 nt for fly piRNAs) including the size markers using a fresh scalpel or razor blade (*see* Fig. 1b, left). Make sure to remove the plastic wrap from the gel slice, and then transfer the gel slice to a low-retention microcentrifuge tube. Add 425 μL of 400 mM NaCl, and elute overnight (at least 8 h) with agitation (*see* **Note 4**).
11. Re-expose the gel to confirm that the fragments were excised correctly.

3.3 Cleanup of Small RNA Samples

1. Centrifuge the samples for 1 min at 20,000 × *g* and transfer the supernatant of each sample to a Micropore 0.22 μm filter microcentrifuge tube (*see* **Note 5**). Then, centrifuge the samples for 2 min at 13,000 × *g*. For each sample, transfer the flow-through containing the eluted RNAs to a low-retention microcentrifuge tube, and add 1.2 mL of 100 % ethanol and 1 μL of GlycoBlue. Incubate the samples at −20 °C for at least 2 h.
2. Precipitate the samples by centrifugation for 30 min at 20,000 × *g* at 4 °C. Wash each pellet with 500 μL of ice-cold 80 % ethanol by centrifugation for 5 min at 20,000 × *g* at 4 °C. Discard the supernatant and air-dry each pellet for 5 min. Resuspend the RNA pellets in 12 μL nuclease-free water by gently flicking each tube several times (do not pipet up and down as pellets are easily lost).

3.4 Ligation of 3′ Linker and Gel Purification of Ligation Product

1. Mix the size selected small RNAs with 2 μL of DMSO, 2 μL of 10× ATP-free T4 RNA ligase buffer, 1 μL of RNase Inhibitor, 1 μL of 100 μM *oLig3* linker (Table 1), and 2 μL of T4 RNA ligase 2, truncated K227Q (*see* **Note 6**). Briefly centrifuge to collect the sample at the bottom of the tube. Incubate the reaction at room temperature for 2 h (*optional: incubate overnight at 16°C*).
2. Prepare a 0.75-mm thick, 10 % denaturing polyacrylamide gel using the UreaGel system. For 50 mL of gel, mix 20 mL of concentrate, 25 mL of diluent, and 5 mL of 10× buffer (all included in the UreaGel system) in a Falcon tube. Add 400 μL of 10 % APS and 20 μL of TEMED, mix and immediately cast the gel. Once polymerized, clean each well and pre-run the gel as previously described in **step 3** in Subheading 3.2.

3. Dilute 1 μL of radioactive ladder in 40 μL 1× gel loading buffer. Add one volume of 2× gel loading buffer to each sample. Next, denature the 3′-ligated small RNAs and ladder by heating for 2 min at 95 °C, and then immediately cool on ice for 2 min. Briefly centrifuge to collect the sample at the bottom of the tube.
4. Following the pre-run, rinse the wells again using a syringe. Immediately load the samples and ladder using gel loading tips. Run the gel at 400 V until the bromophenol blue dye travels three-quarters of the length of the gel (~1.5 h).
5. Disassemble the gel apparatus, while keeping the gel on one glass plate.
6. Pipet 0.25 μL of radioactive ladder onto four small pieces of Whatman paper. Use adhesive tape to attach the radioactive paper pieces onto the edges of the glass plate. Carefully cover the gel and glass plate with plastic wrap.
7. Expose the gel to a phosphor screen for approximately 10 min (time depends on the amount of radioactivity used), and then scan the image with a phosphorimager and print the gel image (select actual size or 1:1 scale).
8. Place the printout under the glass plate and use the radioactive Whatman paper pieces to properly align the gel and printout.
9. For each sample, precisely excise the desired gel fragment (37–46 nt for fly piRNAs) including the size markers using a fresh scalpel or razor blade (*see* Fig. 1b, middle). Make sure to remove the plastic wrap from the gel slice, and then transfer the gel slice into a low-retention microcentrifuge tube. Add 425 μL of 400 mM NaCl, and elute overnight (at least 8 h) with agitation (*see* **Note 4**).
10. Re-expose the gel to confirm that the fragments were excised correctly.
11. Cleanup the RNA samples using the instructions from Subheading 3.3.

3.5 Ligation of 5′ Linker and Gel Purification of Ligation Product

1. Mix the 3′-ligated small RNAs (12 μL) with 2 μL of DMSO, 2 μL of 10× T4 RNA ligase buffer, 1 μL of RNase inhibitor, 1 μL of 100 μM *oLig5* linker (Table 1), and 2 μL of T4 RNA ligase (*see* **Note 6**). Briefly centrifuge to collect the sample at the bottom of the tube. Incubate the reaction at 37 °C for 2 h.
2. Prepare a 0.75-mm thick, 8 % denaturing polyacrylamide gel using the UreaGel system. For 50 mL of gel, mix 16 mL of concentrate, 29 mL of diluent, and 5 mL of 10× buffer (all included in the UreaGel system) in a Falcon tube. Add 400 μL of 10 % APS and 20 μL of TEMED, mix and immediately cast the gel. Once polymerized, clean each well and pre-run the gel as previously described in **step 3** in Subheading 3.2.

3. Dilute 0.5 μL of radioactive ladder in 40 μL 1× gel loading buffer and add one volume of 2× gel loading buffer to each sample. Next, denature the RNA samples and ladder by heating for 2 min at 95 °C, and then cool on ice for 2 min. Briefly centrifuge to collect the sample at the bottom of the tube.
4. Following the pre-run, rinse the wells again using a syringe. Immediately load the samples and ladder using gel loading tips. Run the gel at 400 V until the bromophenol blue dye travels three-quarters of the length of the gel (~1 h).
5. Disassemble the gel apparatus, while keeping the gel on one glass plate.
6. Pipet 0.25 μL of radioactive ladder onto four small pieces of Whatman paper. Use adhesive tape to attach the radioactive paper pieces onto the edges of the glass plate. Carefully cover the gel and glass plate with plastic wrap.
7. Expose the gel to a phosphor screen for approximately 20 min (time depends on amount of radioactivity used), and then scan the image with a phosphorimager and print the gel image (select actual size or 1:1 scale).
8. Place the printout under the glass plate and use the radioactive Whatman paper pieces to properly align the gel and printout.
9. For each sample, precisely excise the desired gel fragment (69–78 nt for fly piRNAs) including the size markers using a fresh scalpel or razor blade (*see* Fig. 1b, right). Make sure to remove the plastic wrap from the gel slice, and then transfer the gel slice into a low-retention microcentrifuge tube. Add 425 μL of 400 mM NaCl, and elute overnight (at least 8 h) with agitation (*see* **Note 4**).
10. Re-expose the gel to confirm that the fragments were excised correctly.
11. Cleanup the RNA samples using the instructions from Subheading 3.3, but resuspend the pellet with 11 μL of nuclease-free water.

3.6 Reverse Transcription

1. Mix 11 μL of ligated RNAs, 1 μL of 10 mM dNTP mix, and 1 μL of 100 μM *oPCR3* primer (Table 1) in a PCR tube.
2. Incubate for 3 min at 65 °C, then immediately transfer the reaction mixture to ice for 5 min.
3. Add 4 μL of 5× First-Strand Buffer, 1 μL of 0.1 M DTT, 1 μL of RNase inhibitor, and 1 μL of SuperScript III reverse transcriptase to the mixture.
4. Mix and briefly centrifuge to collect the sample at the bottom of the tube.
5. Incubate for 60 min at 50 °C, then at 70 °C for 15 min, then at 4 °C until use in the next step.

3.7 PCR Amplification of cDNA

1. Set up a reaction mixture with a total volume of 100 μL in a PCR tube containing 2.5 μL of cDNA (from Subheading 3.6), 1 μL of 100 μM *oPCR5* (Table 1), 1 μL of 100 μM *oPCR3* (Table 1), 10 μL of 10× PCR buffer, 10 μL of 2 mM dNTP mix, 6 μL of $MgSO_4$ (25 mM), 1 μL of KOD Polymerase, and 68.5 μL of nuclease-free water.
2. Mix and briefly centrifuge to collect the sample at the bottom of the tube.
3. Place the tubes in a thermal cycler and run the following PCR program:

   95 °C for 2 min
   followed by five cycles of:
   95 °C for 15 s
   54 °C for 30 s
   72 °C for 15 s
   Run an additional 17 cycles of (*see* **Note** 7):
   95 °C for 15 s
   60 °C for 30 s
   72 °C for 15 s
   followed by:
   72 °C for 2 min
   4 °C forever
4. Transfer the PCR mixture to a microcentrifuge tube, add 170 μL of nuclease-free water, 30 μL of 3 M NaOAc (pH 4.8), and three volumes of 100 % ethanol (900 μL), and mix.
5. Incubate at −20 °C for 30 min.
6. Centrifuge at 20,000 × *g* for 30 min at 4 °C.
7. Decant supernatant and wash pellet with 500 μL of ice-cold 80 % ethanol.
8. Remove all ethanol, air-dry pellet, and then resuspend the pellet in 23.7 μL of nuclease-free water.

3.8 PmeI Digest and Gel Purification of Amplified cDNA

1. Set up a 30 μL restriction digest by mixing 23.7 μL of DNA (from **step 8** in Subheading 3.7) with 3 μL of 10× NEBuffer 4, 0.3 μL of BSA, and 3 μL of PmeI restriction enzyme.
2. Mix and briefly centrifuge to collect the sample at the bottom of the tube.
3. Incubate for 3 h at 37 °C.
4. Mix each sample with 6 μL of 6× DNA loading dye.
5. Prepare a 2 % TAE agarose gel containing 500 ng/mL of EtBr.
6. Load all samples (in every other lane to avoid cross-contamination) and 5 μL of 50 bp ladder.
7. Run the gel at constant 100 V for ~45 min.

8. View the gel with a long-wave UV source. Excise the correct DNA band (~108–117 nt when starting with 19–28 nt sized small RNAs) with a clean razor blade or scalpel and transfer the gel slice to a microcentrifuge tube. To avoid contamination, use a fresh razor blade or scalpel for each sample.
9. Cleanup the sample using Promega PCR purification kit according to the manufacturer's instructions. Elute each library with 30 μL nuclease-free water.
9. Measure the DNA concentration by NanoDrop or another suitable spectrophotometer.
10. Dilute the samples to 10 nM and sequence on the appropriate platform (*see* **Note 2**) or store at −20 °C.

4 Notes

1. The length of the RNA size selection markers can be varied depending on the desired cloning range. For libraries comprised of *Drosophila* piRNAs, typically 19- and 28 nt size markers are used. As murine piRNAs are slightly longer, a cloning range of 19–33 nt is recommended. For miRNA libraries, size selection of 19–24 nt is appropriate. Important: Oligonucleotides must contain a PmeI recognition sequence for later removal of converted cloning markers. While we use PmeI in this protocol, it is also possible to substitute other restriction endonucleases (preferentially 8-base cutters) and make the necessary adjustments in the sequences of the size selection markers. It is recommended to use an octanucleotide-recognizing restriction endonuclease to minimize degradation of the small RNA pools within the library.
2. As sequencing technologies improve and provide more reads per run, the protocol can easily be adapted for multiplexing using custom or Illumina-provided adapters and barcoding primers.
3. Several protocols exist to obtain total RNAs (e.g., TRIzol reagent; Invitrogen). Other methods that extract small RNAs (e.g., mirVANA; Ambion) can be used as well. It is also possible to use RNAs from immunoprecipitates (e.g., immunoprecipitation of Aubergine-bound piRNAs) as starting material. However, most column-based kits remove RNAs smaller than 200 bp and are therefore not suitable for small RNA cloning. RNA concentration should be determined using a NanoDrop or another suitable spectrophotometer. Ensure the highest possible quality of the RNA starting material by Bioanalyzer (Agilent) or PAGE stained with EtBr. Optionally, rRNAs (especially *Drosophila* 2S rRNA) can be depleted by several available kits or other reported methods (e.g., RNaseH digest) prior to size selection or before 3′ ligation.

4. Use of dializers for fragment recovery instead of overnight elution is possible. In brief, insert the gel slice and 450 μL of RNase-free water into a dialyzer tube (D-Tube Dialyzer Midi, MWCO 3.5 kDa; Novagen). Position the dialyzer in an appropriate gel box, and add 1× TAE to allow current flow. Run for 18 min at 100 V, then reverse the poles and run for 2 min to pull samples off the dialysis tubing walls. Transfer all RNase-free water (containing the eluted RNAs) from the dialyzer tube to a fresh RNase-free microcentrifuge tube. Add 1 μL GlycoBlue and 3 volumes of 100 % ethanol. Mix and incubate at −20 °C for at least 2 h. Continue the precipitation as described above (spin, wash, and resuspend).
5. It is recommended to save the gel pieces after transferring the supernatant to a spin column in case of failure or loss of the sample. Store gel slices at −20 °C. To recover RNA from the gel slices, add 425 μL of 400 mM NaCl and incubate at room temperature with agitation for at least 8 h. Then follow the steps outlined in Subheading 3.3 to clean up the samples.
6. T4 RNA ligase 2, truncated K227Q is optimized to ligate the pre-adenylated 5′ terminus of a nucleic acid (RNA or DNA; here the *oLig3* linker) with the 3′ hydroxyl terminus of a small RNA (regardless of the presence of additional 3′ end modifications like 2′-*O*-methyl groups) [15, 16]. In contrast, T4 RNA ligase catalyzes the formation of a phosphodiester bond using a 5′ mono-phosphorylated RNA and a nucleic acid with 3′ hydroxyl end as templates [17].
7. The appropriate number of cycles varies depending on the starting material and loss during the cloning procedure. To avoid over-amplification, it is recommended to use the fewest number of cycles possible.

Acknowledgments

We thank past and current members of the Hannon laboratory that helped improve this protocol, especially Alexei Aravin, Julius Brennecke, and Colin Malone.

References

1. Lee RC, Feinbaum RL, Ambros V (1993) The C. elegans heterochronic gene lin-4 encodes small RNAs with antisense complementarity to lin-14. Cell 75:843–854
2. Wightman B, Ha I, Ruvkun G (1993) Posttranscriptional regulation of the heterochronic gene lin-14 by lin-4 mediates temporal pattern formation in C. elegans. Cell 75:855–862
3. Lagos-Quintana M, Rauhut R, Lendeckel W, Tuschl T (2001) Identification of novel genes coding for small expressed RNAs. Science 294:853–858
4. Lau NC, Lim LP, Weinstein EG, Bartel DP (2001) An abundant class of tiny RNAs with probable regulatory roles in Caenorhabditis elegans. Science 294:858–862

5. Lee RC, Ambros V (2001) An extensive class of small RNAs in Caenorhabditis elegans. Science 294:862–864
6. Aravin A et al (2006) A novel class of small RNAs bind to MILI protein in mouse testes. Nature 442:203–207
7. Girard A, Sachidanandam R, Hannon GJ, Carmell MA (2006) A germline-specific class of small RNAs binds mammalian Piwi proteins. Nature 442:199–202
8. Grivna ST, Beyret E, Wang Z, Lin H (2006) A novel class of small RNAs in mouse spermatogenic cells. Genes Dev 20:1709–1714
9. Lau NC et al (2006) Characterization of the piRNA complex from rat testes. Science 313:363–367
10. Ruby JG et al (2006) Large-scale sequencing reveals 21U-RNAs and additional microRNAs and endogenous siRNAs in C. elegans. Cell 127:1193–1207
11. Brennecke J et al (2007) Discrete small RNA-generating loci as master regulators of transposon activity in Drosophila. Cell 128:1089–1103
12. Gunawardane LS et al (2007) A slicer-mediated mechanism for repeat-associated siRNA 5′ end formation in Drosophila. Science 315:1587–1590
13. Houwing S et al (2007) A role for Piwi and piRNAs in germ cell maintenance and transposon silencing in Zebrafish. Cell 129:69–82
14. Pfeffer S, Lagos-Quintana M, Tuschl T (2005) Cloning of small RNA molecules. Curr Protoc Mol Biol Chapter 26, Unit 26.4, edited by Frederick M. Ausubel … [et al.]
15. Hafner M et al (2008) Identification of microRNAs and other small regulatory RNAs using cDNA library sequencing. Methods 44:3–12
16. Zhuang F, Fuchs RT, Robb GB (2012) Small RNA expression profiling by high-throughput sequencing: implications of enzymatic manipulation. J Nucleic Acids 2012:360358
17. Silber R, Malathi VG, Hurwitz J (1972) Purification and properties of bacteriophage T4-induced RNA ligase. Proc Natl Acad Sci U S A 69:3009–3013

Chapter 17

Analysis of Piwi-Loaded Small RNAs in *Terahymena*

Tomoko Noto, Henriette M. Kurth, and Kazufumi Mochizuki

Abstract

Scan RNAs (scnRNAs) are developmentally regulated siRNAs of ~26–32 nucleotides in length that are involved in programmed DNA elimination in *Tetrahymena*. scnRNAs are loaded onto the Piwi-related protein Twi1p and 2′-*O*-methylated at their 3′ termini. We describe two alternative strategies for analyzing the Twi1p-loaded scnRNAs: preparation of loaded scnRNAs by immuno-purification of the Twi1p–scnRNA complex and exclusion of non-methylated scnRNAs during cDNA library construction using periodate oxidation.

Key words Argonaute, Piwi, siRNA, scnRNA, 2′-*O*-methylation, *Tetrahymena*

1 Introduction

The ciliated protozoan *Tetrahymena* expresses developmentally regulated siRNAs, called scnRNAs that are exclusively detected during conjugation, the sexual reproduction process [1–3]. The Dicer protein Dcl1p is required for the production of scnRNAs [4, 5]. The scnRNA duplex produced by Dcl1p is loaded onto the Argonaute-family protein Twi1p [2]. Although Twi1p binds to siRNA, it is rather closely related to the Piwi-clade proteins of the Argonaute family [1]. The passenger strand of the scnRNA duplex is cleaved by the endoribonuclease activity of Twi1p and discarded [6]. The remaining guide strand is 2′-*O*-methylated by the RNA methyltransferase Hen1p [7]. Twi1p–scnRNA complexes are imported into the macronucleus, where they interact with chromatin by binding nascent transcripts and induce programmed DNA elimination [8]. *Tetrahymena* also constitutively expresses ~23–24 nt siRNAs that are produced via a different biogenesis mechanism [9–11]. For the most part, these constitutively expressed siRNAs are not 2′-*O*-methylated [12].

Here, we describe two alternative methods for preparing cDNA libraries of Twi1p-loaded scnRNAs. The first method includes preparation of loaded scnRNAs by immuno-purification

Mikiko C. Siomi (ed.), *PIWI-Interacting RNAs: Methods and Protocols*, Methods in Molecular Biology, vol. 1093,
DOI 10.1007/978-1-62703-694-8_17, © Springer Science+Business Media, LLC 2014

of Twi1p–scnRNA complexes. The second method excludes unloaded scnRNAs from 3′-adopter ligation during cDNA library construction by periodate oxidation. The latter strategy is based on the fact that 2′-*O*-methylation of scnRNA at its 3′-terminus, which occurs only on the Twi1p-loaded scnRNAs, prevents the formation of 2′,3′-dialdehydes by periodate treatment. These two strategies are both valid ways to concentrate the Twi1p-loaded scnRNAs, though the level of contamination by rRNA fragments is lower in the oxidation method. Therefore, the oxidation method is a convenient way to prepare cDNA libraries from samples in which 2′-*O*-methylation of scnRNAs has been confirmed. On the other hand, the immunoprecipitation method can be used with samples in which the methylation state of the scnRNAs is unknown.

2 Materials

2.1 Preparation of Anti-Twi1p Antibody Beads

1. EZview™ Red Protein A Affinity Gel (Sigma).
2. 1.5 mL low-binding tubes (Sorenson).
3. IP wash buffer: 50 mM Tris–HCl (pH 7.5), 150 mM NaCl, 2 mM $MgCl_2$.
4. IP wash buffer including 0.1 % BSA: 50 mM Tris–HCl (pH 7.5), 150 mM NaCl, 2 mM $MgCl_2$, 0.1 % BSA (Albumin Fraction V).
5. Anti-Twi1p antibody (*see* **Note 1**).

2.2 Preparation of Cell Lysate

1. Mating *Tetrahymena thermophila* cells (*see* **Note 2**).
2. IP buffer: IP wash buffer plus 1× Complete Protease Inhibitor (EDTA-free) (Roche).
3. RiboLock™ RNase Inhibitor (Fermentas).
4. Sonicator, Omni-Ruptor 250 (Omni) equipped with an OR-T-156 probe (Omni).
5. 1.5-mL low-binding tubes (Sorenson).

2.3 Immunoprecipitation and RNA Isolation

1. Trizol reagent (Life technologies).
2. 1.5-mL low-binding tubes (Sorenson).
3. GlycoBlue (Life technologies).
4. Formamide (*see* **Note 3**).

2.4 RNA Preparation for Periodate Oxidation Reaction

1. Trizol reagent (Life technologies).
2. 1.5-mL low-binding tubes (Sorenson).

2.5 Periodate Oxidation

1. Borax buffer: 4.375 mM borax (anhydrous), 50 mM boric acid (pH 8.6); store at RT in the dark.
2. 0.2 M sodium periodate; store at RT in the dark.
3. 5 M NaCl.
4. Formamide (*see* **Note 3**).

2.6 Preparation of ^{32}P-Labeled Size Markers

1. Decade marker (Life technologies).
2. Size-marker oligo RNAs, 26 and 32 nt (*see* **Note 4**):
   5′-GUCGUACGCGGAUA<u>GUUUAAAC</u>UGU-3′;
   5′-AUCUUGGUCGUACGCGGAAU<u>GUUUAAAC</u>UGU-3′
   (PmeI sites underlined).
3. [γ-^{32}P] ATP, 6,000 Ci/mmol (Perkin Elmer).
4. T4 polynucleotide kinase, 10 U/μL (Fermentas).
5. Liquid scintillation counter, Triathler multilabel tester (Hidex).
6. Mini Quick Spin RNA Columns (Roche).

2.7 Purification of scnRNAs

1. SequaGel system (National Diagnostics).
2. 10× SeqGel loading dye: 0.25 % Xylene cyanol, 0.25 % bromophenol blue, 5 mM EDTA.
3. X-ray film (Amersham Hyperfilm™ MP, GE Healthcare).
4. 1.5-mL low-binding tubes (Sorenson).
5. 0.4 M NaCl.
6. Eppendorf Thermomixer comfort (Eppendorf).
7. GlycoBlue (Life technologies).
8. Ultrafree-MC, 0.22 μm (Millipore).
9. 80 % ethanol, ice cold.

2.8 Ligation of 3′ Linker

1. T4 RNA ligase 2, truncated K227Q, 200 U/μL (New England Biolabs).
2. 10× ATP-free T4 RNA ligase buffer (supplied with **item 1** in Subheading 2.8).
3. 50 % polyethylene glycol (PEG) 8000 (supplied with **item 1** in Subheading 2.8).
4. 3′ DNA linker (Integrated DNA Technologies):
   5′-App-TCGTATGCCGTCTTCTGCTTG-ddC-3′.
5. SequaGel system (National Diagnostics).
6. 10× SeqGel loading dye: 0.25 % Xylene cyanol, 0.25 % bromophenol blue, 5 mM EDTA.
7. X-ray film (Amersham Hyperfilm™ MP, GE Healthcare).
8. 1.5-mL low-binding tubes (Sorenson).

9. 0.4 M NaCl.
10. Eppendorf Thermomixer comfort (Eppendorf).
11. GlycoBlue (Life technologies).
12. Ultrafree-MC, 0.22 μm (Millipore).
13. 80 % ethanol, ice cold.

2.9 Ligation of 5′ Linker

1. T4 RNA ligase, 5 U/μL (Life technologies).
2. 5′ RNA linker, HPLC purified: 5′-GUUCAGAGUUCUACAGUCCGACGAUC-3′.

2.10 Reverse Transcription

1. Sol-rev primer, HPLC purified: 5′-CAAGCAGAAGACGGCATACGA-3′.
2. Superscript II Reverse Transcriptase (Life technologies).
3. 10 mM each dNTP mix.
4. 0.7-mL low-binding tubes (Sorenson).

2.11 PCR Amplification of cDNA

1. 0.2-mL PCR tubes.
2. Taq DNA polymerase, 5 U/μL (5-PRIME).
3. 2 mM each dNTP mix.
4. Sol-fw primer, HPLC purified: 5′-AATGATACGGCGACCACCGACAGGTTCAGAGTT CTACAGTCCGA-3′.
5. Sol-rev primer, HPLC purified: 5′-CAAGCAGAAGACGGCATACGA-3′.
6. 1.5-mL low-binding tubes (Sorenson).
7. 0.5 M NaCl.
8. Phenol/chloroform/isoamyl alcohol (25:24:1).
9. GlycoBlue (Life technologies).
10. 70 % ethanol, ice cold.

2.12 Purification of cDNA Library

1. PmeI (New England Biolabs).
2. 10 mg/mL BSA.
3. 0.5 M NaCl.
4. GlycoBlue (Life technologies).
5. 70 % ethanol, ice cold.
6. 2 % NuSieve GTG Agarose (Lonza) gel.
7. GeneRuler™ DNA Ladder Mix (Fermentas).
8. 1 μg/mL ethidium bromide solution.
9. 1.5-mL low-binding tubes (Sorenson).
10. QIAquick Gel Extraction kit (Qiagen).

3 Methods

3.1 Preparation of Anti-Twi1p Antibody Beads

1. Place 200 μL of EZview™ Red Protein A Affinity Gel (1:1 suspension, bed volume = 100 μL) in a 1.5-mL low-binding tube (*see* **Note 5**).
2. Wash beads three times with 1 mL of IP wash buffer including 0.1 % BSA.
3. Resuspend beads in 1 mL of IP wash buffer including 0.1 % BSA and 80 μL of anti-Twi1p antibody.
4. Rotate at 4 °C for 2 h.
5. Spin down at 1,000 × *g* for 10 s and remove the supernatant.
6. Add 1 mL of IP wash buffer including 0.1 % BSA.
7. Repeat **steps 5** and **6** five more times.
8. Proceed to Subheading 3.3 or store beads at 4 °C.

3.2 Preparation of Cell Lysate

1. Spin down 5×10^6 mating cells (e.g., 10 mL of a 5×10^5 cell/mL suspension) in a 15 mL conical tube and remove the supernatant (*see* **Note 6**).
2. Resuspend the cells in 2.5 mL of IP buffer including 25 μL of RiboLock™ RNase Inhibitor.
3. Sonicate (30 pulses) using an Omni-Ruptor 250 with an OR-T-156 probe set to 20 % power and 50 % pulse (*see* **Note 7**).
4. Transfer the lysate to two 1.5-mL low-binding tubes and centrifuge at 18,000 × *g* for 20 min at 4 °C.
5. Collect 1 mL of supernatant and transfer to a fresh 1.5-mL low-binding tube (*see* **Note 8**).

3.3 Immunoprecipitation and RNA Isolation

1. Remove buffer from the beads prepared in Subheading 3.1.
2. Resuspend the beads in 1 mL of the cell lysate from Subheading 3.2.
3. Rotate for 3 h at 4 °C.
4. Spin down at 1,000 × *g* for 10 s and remove the supernatant.
5. Add 1 mL of IP wash buffer.
6. Repeat **steps 4** and **5** five more times.
7. Remove the supernatant and add 1 mL of Trizol.
8. Vortex for 30 s and let the tube stand for 5 min at RT.
9. Spin down at 18,000 × *g* for 1 min at 4 °C.
10. Collect the supernatant in a fresh 1.5-mL low-binding tube.
11. Add 200 μL of chloroform to the supernatant, mix vigorously for 1 min, and let the tube stand for 2 min at RT.
12. Centrifuge at 18,000 × *g* for 20 min at 4 °C.

13. Transfer 500 μL of the upper aqueous phase to a fresh 1.5-mL low-binding tube.
14. Add 1.3 μL of GlycoBlue to the aqueous phase and vortex briefly.
15. Add 500 μL of isopropanol and mix well.
16. Incubate at 20 °C for >30 min.
17. Centrifuge at 18,000 × *g* for 30 min at 4 °C.
18. Remove the supernatant and let the pellet dry on the bench top for 3 min (*see* **Note 9**).
19. Dissolve the pellet in 20 μL of formamide.
20. Proceed to Subheading 3.7 or store the sample at −80 °C until use.

3.4 RNA Preparation for Periodate Oxidation Reaction

1. Prepare and collect cells as in Subheading 3.2.
2. Lyse cells in 1 mL of Trizol regent, transfer to a fresh 1.5-mL tube, and let the tube stand at RT for 5 min.
3. Proceed to **step 4** or store the sample at −80 °C.
4. Add 200 μL of chloroform and mix vigorously for 1 min.
5. Centrifuge at 18,000 × *g* for 20 min at 4 °C.
6. Carefully transfer 500 μL of the upper aqueous phase to a fresh 1.5-mL low-binding tube.
7. Add 500 μL of isopropanol and mix well.
8. Let the tube stand at RT for 5 min.
9. Centrifuge at 18,000 × *g* for 20 min at 4 °C.
10. Remove the supernatant and let the pellet dry on the bench top for 3 min (*see* **Note 9**).
11. Repeat **steps 2–10** and proceed to Subheading 3.5 (*see* **Note 10**).

3.5 Periodate Oxidation

1. Dissolve the pellet in 35 μL of Borax buffer.
2. Add 5 μL of 0.2 M sodium periodate and mix well.
3. Incubate for 10 min at RT in the dark.
4. Add 4 μL of glycerol and mix well.
5. Incubate for 10 min at RT in the dark.
6. Add 56 μL of nuclease-free water and 4 μL of 5 M NaCl; mix well.
7. Add 250 μL of ethanol and mix well.
8. Incubate for >30 min at −20 °C.
9. Centrifuge at 18,000 × *g* for 30 min at 4 °C.
10. Remove the supernatant and let the pellet dry on the bench top for 3 min.

11. Dissolve the pellet in 20 μL of formamide.
12. Proceed to Subheading 3.7 or store the sample at −80 °C until use.

3.6 Preparation of ^{32}P-Labeled Size Markers

1. Prepare the Decade marker according to the manufacturer's protocol.
2. Set up the following reaction:

5 μM 26 or 32-nt oligo RNA	2 μL
10× T4 Polynucleotide Kinase Buffer	2 μL
[γ-^{32}P] ATP	1.5 μL
T4 Polynucleotide Kinase (10 U/μL)	1 μL
Nuclease-free water	13.5 μL

3. Incubate for ~30–60 min at 37 °C.
4. Add 10 μL of nuclease-free water.
5. Apply the reaction to a Mini Quick Spin RNA Column.
6. Centrifuge at 1,000 × *g* for 4 min at RT.
7. Collect the flow-through, which contains the labeled RNA oligo.
8. Add 1 μL of labeled RNA oligo or Decade marker to a 1.5-mL tube and measure radioactivity using liquid scintillation counter in Cerenkov counting (^{3}H) mode.

3.7 Purification of scnRNAs

1. Using the SequaGel system, prepare a 15 % polyacrylamide/urea gel of 16 × 16 cm in size and 1 mm in thickness with combs of 6 mm in width.
2. Pre-run the gel for 1 h at constant voltage (250 V).
3. While pre-running the gel, set up the following samples:

RNA in formamide (from Subheading 3.3 or Subheading 3.5)	20 μL
[γ-^{32}P] ATP labeled 26-nt RNA oligo	20,000 cpm
[γ-^{32}P] ATP labeled 32-nt RNA oligo	20,000 cpm
10× SeqGel loading dye	4 μL
Nuclease-free water	up to 40 μL

4. Denature the RNA sample and the labeled decade marker (**step 1** in Subheading 3.6) by incubating for 5 min at 95 °C followed by chilling on ice for >2 min.
5. Load 100,000 cpm of the labeled decade marker into a lane.
6. Load the RNA sample into another lane.

7. Run the gel at constant wattage (15–20 W) for 1–2 h until BPB reaches the middle of the gel.
8. Remove the gel from glass plate and place on a used X-ray film.
9. Cover the gel with plastic wrap and expose to an X-ray film with enhancer screen for 1 h at RT.
10. Develop the film and mark the gel region containing the two RNA oligos on the plastic wrap (*see* **Note 11**).
11. Cut out the marked gel region using a clean razor.
12. Put the gel pieces into a 1.5 mL tube and spin down briefly.
13. Smash the gel pieces with an end-closed 1 mL pipette tip (*see* **Note 12**).
14. Add 400 μL of 0.4 M NaCl.
15. Snap-freeze in liquid nitrogen and incubate overnight at RT with agitation at 1,300 rpm in an Eppendorf Thermomixer.
16. Apply the gel homogenate to an Ultrafree-MC column using a 1 mL pipette tip with its end cut off (*see* **Note 13**).
17. Centrifuge at 10,000 × *g* for 1 min at RT.
18. Collect the flow-through into a fresh 1.5-mL low-binding tube.
19. Add 1.3 μL of GlycoBlue and 1 mL of ethanol.
20. Incubate for ~3–6 h at −20 °C.
21. Centrifuge at 18,000× *g* for 30 min at 4 °C.
22. Remove the supernatant and add 1 mL of 80 % ethanol.
23. Spin down at 18,000× *g* for 1 min and remove supernatant.
24. Let the pellet dry on the bench top for 3 min.
25. Dissolve the pellet in 9 μL of nuclease-free water.

3.8 Ligation of 3′ Linker

1. Set up the following reaction (total reaction volume = 20 μL):

Gel-purified RNA from Subheading 3.7	9 μL
10× ATP-free T4 RNA ligase buffer	2 μL
50 % PEG 8000	6 μL
50 μM 3′ DNA linker	1 μL
T4 RNA ligase 2, truncated K227Q	2 μL

2. Incubate for 30 min at RT.
3. Incubate overnight at 4 °C.

4. Using the SequaGel system, prepare a 15 % polyacrylamide/urea gel of 16×16 cm in size and 1 mm in thickness, with combs of 6 mm in width.
5. Pre-run the gel for 1 h at constant voltage (250 V).
6. Add 20 μL of formamide and 4 μL of 10× SeqGel loading dye to the reaction.
7. Heat for 5 min at 95 °C for inactivation.
8. Chill the sample and marker on ice for >2 min.
9. Load 100,000 cpm of the labeled decade marker into a lane.
10. Load all of the ligated RNA into another lane.
11. Run the gel at constant wattage (15–20 W) for ~1–2 h (until BPB reaches the bottom of the gel).
12. Remove the gel from glass plate and place on a used exposed film.
13. Cover the gel with plastic wrap and expose to an X-ray film with enhancer screen for 1 h at RT.
14. Develop the film and mark the gel region containing the two linker-ligated RNA oligos on the plastic wrap (*see* **Note 14**).
15. Cut out the marked gel region using a clean razor.
16. Put the gel pieces into a fresh low-binding 1.5-mL tube and spin down briefly.
17. Smash the gel pieces with an end-closed (*see* **Note 12**) 1-mL pipette tip.
18. Add 400 μL of 0.4 M NaCl.
19. Snap-freeze in liquid nitrogen and incubate overnight at RT with agitation at 1,300 rpm.
20. Apply the gel homogenate to an Ultrafree-MC column using a 1-mL pipette tip with its end cut off (*see* **Note 13**).
21. Centrifuge at 10,000×*g* for 1 min at RT.
22. Collect the flow-through into a fresh low-binding 1.5-mL tube.
23. Add 1.3 μL of GlycoBlue and 1 mL of ethanol; mix well.
24. Incubate for ~3–6 h at −20 °C.
25. Centrifuge at 18,000×*g* for 30 min at 4 °C.
26. Remove the supernatant and add 1 mL of 80 % ethanol.
27. Spin down at 18,000×*g* for 1 min and remove the supernatant.
28. Let the pellet dry on the bench top for 3 min.
29. Dissolve the pellet in 10 μL of nuclease-free water.

3.9 Ligation of 5′ Linker

1. Set up the following reaction (total reaction volume = 20 μL):

Ligated RNA from Subheading 3.8	10 μL
10× T4 RNA ligase buffer	2 μL
30 % PEG 8000	5 μL
50 μM 5′ RNA linker	1 μL
T4 RNA ligase	2 μL

2. Repeat **steps 2–28** in Subheading 3.8.
3. Dissolve the pellet in 6 μL of nuclease-free water.

3.10 Reverse Transcription

1. Set up the following reaction:

Ligated RNA	6 μL
5 μM Sol rev primer	2 μL

2. Incubate for 2 min at 72 °C.
3. Spin down briefly and chill on ice for 2 min.
4. Add the following:

5× First Strand buffer	4 μL
10 mM dNTP mix	2 μL
0.1 M DTT	2 μL

5. Split the sample into two 0.7-mL low-binding tubes (9 μL each).
6. Add either 1 μL of Superscript II Reverse Transcriptase ("+RT" sample) or 1 μL of nuclease-free water ("–RT" control).
7. Incubate for 1 h at 42 °C and then for 15 min at 70 °C.
8. Proceed to Subheading 3.11 or store the sample at –20 °C until use.

3.11 PCR Amplification of cDNA

1. Set up the following reaction (total volume = 100 μL):

First strand cDNA or -RT control from Subheading 3.10	2 μL
10× Taq buffer	10 μL
2 mM dNTP mix	10 μL
100 μM Sol-fw primer	1 μL
100 μM Sol-rev primer	1 μL
MilliQ water	75 μL
Taq DNA polymerase (5 U/μl)	1 μL

2. Perform PCR with the following cycle:

   95 °C for 30 s

   55 °C for 30 s

   72 °C for 60 s

3. Remove 10 μL of reaction to fresh tubes after 8, 10, 12, 14, 16, 18, 20, and 22 cycles; keep on ice.
4. Check the product by gel electrophoresis with 2 % agarose gel and choose the minimum cycle number at which the PCR products (~100 bp) become visible.
5. Set up the following reaction (total volume = 300 μL):

First strand cDNA from Subheading 3.10	6 μL
10× Taq buffer	30 μL
2 mM dNTP mix	30 μL
100 μM Sol-fw primer	3 μL
100 μM Sol-rev primer	3 μL
MilliQ water	225 μL
Taq DNA polymerase	3 μL

6. Aliquot 50 μL of each reaction into six 0.2-mL PCR tubes.
7. Perform PCR with the chosen number of cycles.
8. Collect the PCR product (300 μL) into a fresh 1.5-mL low-binding tube.
9. Transfer half of the product (150 μL) into a fresh 1.5-mL low-binding tube and store the remaining product at −20 °C as a backup.
10. Add 300 μL of 0.5 M NaCl and 450 μL of phenol/chloroform/isoamyl alcohol (pH 8.0).
11. Mix vigorously and centrifuge at 18,000 × *g* for 15 min at RT.
12. Collect ~400 μL of the upper aqueous phase in a fresh 1.5-mL low-binding tube.
13. Add 1.3 μL of GlycoBlue and ~1,000 μL (2.5 times the volume of the collected upper phase) of ethanol; mix well.
14. Incubate for ~3–6 h at −20 °C.
15. Centrifuge at 18,000 × *g* for 30 min at 4 °C.
16. Remove the supernatant and add 1 mL of 70 % EtOH.
17. Spin down at 18,000 × *g* for 1 min and remove supernatant.

18. Let the pellet dry on the bench top for >5 min.
19. Dissolve the pellet in 50.4 μL of MilliQ water.
20. Proceed to Subheading 3.12 or store the sample at −20 °C until use.

3.12 Purification of cDNA Library

1. Set up the following reaction (total volume = 60 μL):

PCR product from **step 19** in Subheading 3.11	50.4 μL
10× NEB buffer 4	6 μL
PmeI, 10,000 U/mL	3 μL
BSA, 10 mg/mL	0.6 μL

2. Incubate for >3 h at 37 °C.
3. Add 390 μL of 0.5 M NaCl and 450 μL of phenol/chloroform/isoamyl alcohol (pH 8.0).
4. Mix vigorously and centrifuge at 18,000 × *g* for 15 min at RT.
5. Collect upper aqueous phase in a fresh tube.
6. Add 1.3 μL of GlycoBlue and 2.5 volumes of ethanol; mix well.
7. Incubate for 3–6 h at −20 °C.
8. Centrifuge at 18,000 × *g* for 30 min at 4 °C.
9. Remove the supernatant and add 1 mL of 70 % EtOH to the pellet.
10. Spin down and remove the supernatant.
11. Air dry for 3 min at RT.
12. Dissolve the pellet in 10 μL of MilliQ water.
13. Prepare 2 % NuSieve GTG Agarose gel (*see* **Note 15**).
14. Load the entire digested PCR product into one lane and 3 μL of DNA marker into another lane.
15. Run the gel at constant voltage (100 V) until BPB reaches 2/3 of the gel length.
16. Stain the gel in 1 μg/mL ethidium bromide solution for 20 min.
17. Cut out the cDNA band (~100 bp) with a clean razor under long-wave UV illumination.
18. Place the gel slice into a pre-weighed 1.5-mL low-binding tube.
19. Weigh the gel slice.

20. Add a volume of QG buffer corresponding to three times the weight of the gel slice.
21. Incubate for 10 min at 37 °C to dissolve the gel slice.
22. Apply to a column from the QIAquick Gel Extraction kit.
23. Centrifuge at 18,000×*g* for 1 min at RT and discard the flow-through.
24. Add 500 μL of QG buffer.
25. Centrifuge at 18,000×*g* for 1 min at RT and discard the flow-through.
26. Apply 750 μL of PE buffer.
27. Centrifuge at 18,000×*g* for 1 min at RT and discard the flow-through.
28. Repeat **steps 26** and **27**.
29. Centrifuge at 18,000×*g* for 5 min at RT.
30. Place the column in a fresh 1.5-mL low-binding tube.
31. Apply 25 μL of MilliQ water.
32. Incubate for 1 min at RT.
33. Centrifuge at 18,000×*g* for 2 min at RT.
34. Collect elution as the cDNA library for scnRNAs.
35. Sequence DNA using the Illumina GAII or HiSeq platform (*see* **Notes 16** and **17**).

4 Notes

1. The anti-Twi1p antibody was raised in rabbit against the peptide KLNKLGRKYFDPASRKEYNEC.
2. Detailed protocols for *Tetrahymena* cell culture and mating induction can be found in refs. 13 and 14.
3. Oxidized formamide can degrade RNA. Aliquots should be made from a freshly opened formamide bottle and should be stored at −20 °C.
4. Because the majority of scnRNAs are 27–30 nt in length, while the other siRNAs present in *Tetrahymena* are 23–24 nt in length, 26- and 32-nt RNA oligos are used as size markers.
5. Use 100 μL bed volume of EZview™ Red Protein A Affinity Gel per immunoprecipitation reaction.

6. Twi1p and scnRNAs are expressed only during mating [1]. Loading of scnRNAs to Twi1p starts as early as 2 h post-mixing and is completed at ~4.5 h post-mixing [6, 15].
7. Sonication conditions must be optimized for each individual sonicator.
8. It is possible to store the cell lysate at −80 °C after snap freezing in liquid nitrogen. However, immunoprecipitation from freshly prepared lysate gives better results.
9. Do not dry the RNA pellet for longer than 3 min. If the RNA pellet is dried completely, it becomes very difficult to dissolve.
10. A second Trizol extraction step is necessary because trace amounts of RNase still remain in the sample after a single Trizol extraction.
11. scnRNAs migrate between the two labeled RNA oligo markers.
12. An end-closed 1-mL pipette tip can be made by deforming the narrow end of a tip using a frame.
13. Cut off ~5 mm of the narrow end of a tip with a clean razor blade.
14. Linker-ligated scnRNAs migrate between the two linker-ligated labeled oligo markers.
15. Although DNA can be eluted from any standard agarose gel using the QIAquick Gel Extraction kit, DNA extracted from NuSieve GTG low-melt agarose gives better sequencing results.
16. The use of mixtures containing a very high fraction (up to 95 %) of 5′ U scnRNAs sometimes disturbs the cluster recognition process during sequencing. Therefore, before sequencing, we add a phiX174 phage DNA library to the cDNA library in a ratio of ~1:10.
17. scnRNA sequences obtained using either of these two library preparation methods similarly map the micronuclear genome (Fig. 1). However, we normally detect many fewer RNAs derived from rRNAs or tRNAs in samples prepared using the oxidation method than in the samples prepared via anti-Twi1p immunoprecipitation (Fig. 2). Therefore, preparation and sequencing of oxidation-resistant small RNAs is a simple and valid method for analyzing Twi1p-loaded scnRNAs that are committed to the DNA elimination pathway, with the caveat that 2′-*O*-methylation of scnRNAs must be confirmed under corresponding experimental conditions.

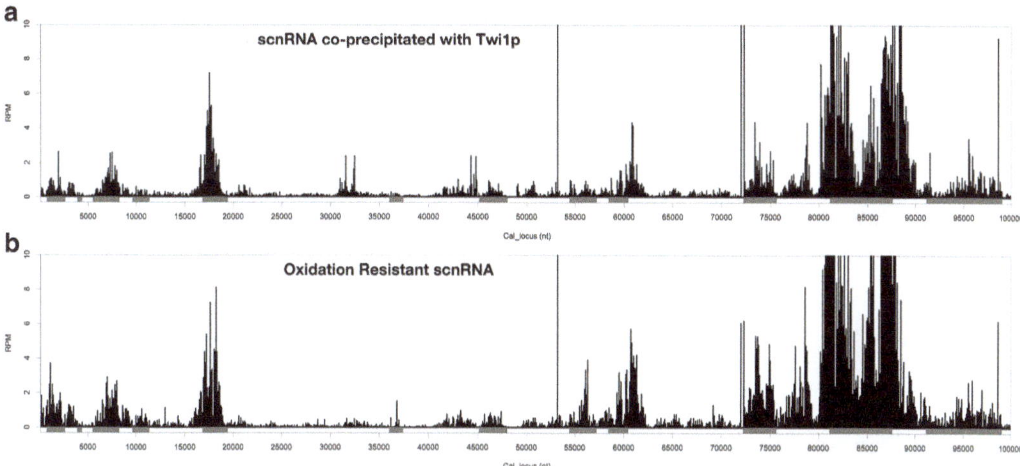

Fig. 1 Comparison of scnRNA sequences obtained from the two different preparation methods. The micronuclear Cal-locus was mapped based on the sequences of the scnRNAs co-immunoprecipitated with Twi1p (**a**) or the oxidation-resistant scnRNAs (**b**). The number (reads per million scnRNA reads [RPM]) of scnRNAs mapping to each position of the locus was divided by its sequence frequency in the draft micronuclear genome. The positions of the internal eliminated sequences (IESs) are marked in *red*. *See* Schoeberl et al. (2012) [15] for the detailed sequence data processing methods

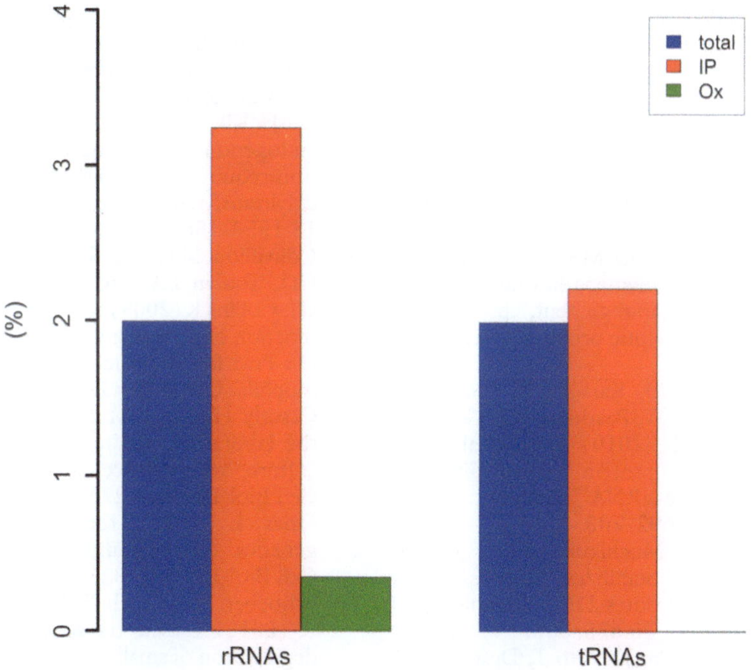

Fig. 2 Sequence reads corresponding to rRNAs or tRNAs in cDNA libraries prepared using the two different methods. The fractions of sequence reads corresponding to rRNAs (*left*) or tRNAs (*right*) in DNA sequences obtained from cDNA libraries for scnRNAs prepared from total RNA (without oxidation, "total", *blue*), RNA co-precipitated with Twi1p ("IP", *red*), and oxidated total RNAs ("Ox", *green*) were compared. Values are expressed as the percentage of sequence reads of 26–32 nt small RNAs

Acknowledgments

We thank Julius Brennecke and his laboratory for providing a small RNA sequencing protocol and for technical advice. Solexa sequencings were performed at the Next Generation Sequencing (NGS) unit of the Campus Science Support Facilities (CSF) (http://csf.ac.at). Our research was supported by an ERC Starting Grant (204986) under the European Community's Seventh Framework Programme, by the Special Research Program (SFB) "RNA regulation of the transcriptome" (F4307-B09) from the Austrian Science Fund (FWF), and by core funding from the Austrian Academy of Sciences to KM.

References

1. Mochizuki K, Fine NA, Fujisawa T, Gorovsky MA (2002) Analysis of a piwi-related gene implicates small RNAs in genome rearrangement in Tetrahymena. Cell 110:689–699
2. Mochizuki K, Gorovsky MA (2004) Conjugation-specific small RNAs in Tetrahymena have predicted properties of scan (scn) RNAs involved in genome rearrangement. Genes Dev 18:2068–2073
3. Lee SR, Collins K (2006) Two classes of endogenous small RNAs in Tetrahymena thermophila. Genes Dev 20:28–33
4. Malone CD, Anderson AM, Motl JA, Rexer CH, Chalker DL (2005) Germ line transcripts are processed by a Dicer-like protein that is essential for developmentally programmed genome rearrangements of Tetrahymena thermophila. Mol Cell Biol 25:9151–9164
5. Mochizuki K, Gorovsky MA (2005) A Dicer-like protein in Tetrahymena has distinct functions in genome rearrangement, chromosome segregation, and meiotic prophase. Genes Dev 19:77–89
6. Noto T, Kurth HM, Kataoka K, Aronica L, Desouza LV, Siu KW, Pearlman RE, Gorovsky MA, Mochizuki K (2010) The Tetrahymena argonaute-binding protein Giw1p directs a mature argonaute-siRNA complex to the nucleus. Cell 140:692–703
7. Kurth HM, Mochizuki K (2009) 2′-O-methylation stabilizes Piwi-associated small RNAs and ensures DNA elimination in Tetrahymena. RNA 15:675–685
8. Aronica L, Bednenko J, Noto T, Desouza LV, Siu KW, Loidl J, Pearlman RE, Gorovsky MA, Mochizuki K (2008) Study of an RNA helicase implicates small RNA-noncoding RNA interactions in programmed DNA elimination in Tetrahymena. Genes Dev 22:2228–2241
9. Lee SR, Collins K (2007) Physical and functional coupling of RNA-dependent RNA polymerase and Dicer in the biogenesis of endogenous siRNAs. Nat Struct Mol Biol 14:604–610
10. Lee SR, Talsky KB, Collins K (2009) A single RNA-dependent RNA polymerase assembles with mutually exclusive nucleotidyl transferase subunits to direct different pathways of small RNA biogenesis. RNA 15:1363–1374
11. Talsky KB, Collins K (2012) Strand-asymmetric endogenous Tetrahymena small RNA production requires a previously uncharacterized uridylyltransferase protein partner. RNA 18:1553–1562
12. Couvillion MT, Lee SR, Hogstad B, Malone CD, Tonkin LA, Sachidanandam R, Hannon GJ, Collins K (2009) Sequence, biogenesis, and function of diverse small RNA classes bound to the Piwi family proteins of Tetrahymena thermophila. Genes Dev 23:2016–2032
13. Cassidy-Hanley DM (2012) Tetrahymena in the laboratory: strain resources, methods for culture, maintenance, and storage. Methods Cell Biol 109:237–276
14. Orias E (2012) Tetrahymena thermophila genetics: concepts and applications. Methods Cell Biol 109:301–325
15. Schoeberl UE, Kurth HM, Noto T, Mochizuki K (2012) Biased transcription and selective degradation of small RNAs shape the pattern of DNA elimination in Tetrahymena. Genes Dev 26:1729–1742

Chapter 18

Effective Gene Knockdown in the *Drosophila* Germline by Artificial miRNA-Mimicking siRNAs

Hailong Wang, Haidong Huang, and Dahua Chen

Abstract

Gene knockdown using double-stranded RNA (dsRNA) is a powerful tool to characterize gene function in *Drosophila*. The *Drosophila* germline provides an elegant model to study the regulation of numerous processes, such as stem cell fate, gametogenesis, piRNA, and piRNA-related gene functions. However, for unknown reasons, traditional dsRNA gene knockdown has not been successful in the germline. Here, we establish a simple gene silencing method for the *Drosophila* germline by the generation of artificial miRNA-mimicking siRNAs. This method, which is different from that of classical dsRNA, mimics natural miRNA biogenesis and enables the analysis of gene functions specifically in the *Drosophila* germline.

Key words miRNA, siRNA, Knockdown, Gene silencing, Germline

1 Introduction

In 1998, Fire and Mello discovered that double-stranded RNAs (dsRNAs) trigger posttranscriptional-gene silencing (PTGS) in *Caenorhabditis elegans* [1]. This discovery initiated the new technology of RNA interference (RNAi) to determine gene function via specific silencing; RNAi is now one of the most popular and efficient tools to study gene function. Long dsRNAs are cleaved into 21–23 nt small RNAs, named small interfering RNAs (siRNAs) [2–4], which are the final effectors loaded onto the RNA-induced silencing complex (RISC) to achieve RNAi. Another small RNA group consists of microRNAs (miRNAs) that are ~23 nt long and play a significant role in posttranscriptional regulation and are necessary for developmental control. These two different small RNAs have distinct biogenesis pathways in Drosophila. siRNAs are processed by Dicer2 and miRNAs are processed by Dicer1. Two research groups have achieved RNAi using artificial miRNAs/siRNAs that mimic the natural microRNA biogenesis pathway [5, 6]. These studies shed a light on how to perform RNAi using endogenous pathways.

Mikiko C. Siomi (ed.), *PIWI-Interacting RNAs: Methods and Protocols*, Methods in Molecular Biology, vol. 1093, DOI 10.1007/978-1-62703-694-8_18, © Springer Science+Business Media, LLC 2014

The *Drosophila* germline system is an excellent system for studying many biological processes, such as gametogenesis, stem cell behavior, germ cell meiosis, and other developmental regulation mechanisms. Studying loss-of-function is one of the best approaches to determine a gene's role, and the roles of many genes have been unveiled by large-scale genetic screening approaches using ethyl methanesulfonate (EMS) [7, 8] and P-element mutagenesis [9, 10]. However, analysis of a gene's role mainly depends on phenotype analysis; therefore, the effects of many genes have not yet been determined because of loss of function lethality. Consequently, given this limitation of traditional forward genetics, it is necessary to develop other ways to remove a gene's function in a tissue-specific manner to bypass lethality.

After the discovery of RNAi, many long dsRNA screens were applied to *Drosophila* and other metazoans to discover new gene function or to expand the known role of a gene [11, 12]. Compared with the traditional time-consuming screening protocols, dsRNA can target any gene product in a rapid and highly efficient way. Moreover, combined with the powerful UAS/Gal4 system, gene knockdown can be performed in a tissue-specific manner. However, dsRNA-mediated gene silencing is ineffective in the female germline because of an unknown reason; therefore, obstacles still exist in the study of germline-specific genes, for example piwi or other piRNA-related genes [13, 14]. Also, mutants of genes that function in the germline, often cause sterility, which masks their real roles. To achieve RNAi specifically in the germline, we asked whether there was any other way to knockdown gene expression without the introduction of dsRNAs. A natural RNAi method had previously been applied in somatic fly cells [15]; therefore, we applied artificial miRNAs/siRNAs to the germline to establish an effective gene silencing method. This method enables the knockdown of a gene of interest in the germline, and provides a powerful tool for systematic reverse genetic screening enabling the identification of hitherto undiscovered genes involved in GSC regulation and germline development.

Many studies have shown differences between miRNA and siRNA biogenesis pathways, in which miRNAs prefer to load into Ago1 and siRNAs assemble on Ago2 [16] (reviewed in [17, 18]). However, Ago1- and Ago2-loading compete with each other; the final loading results depend on the mature RNA duplex and its intrinsic structure [19]. A duplex containing more mismatches preferentially loads onto Ago1, while a duplex with fewer mismatches preferentially loads onto Ago2. The artificial siRNA in this protocol mimics *dme-miR-1*, which only has two mismatches, and the mature product is perfectly complementary to the target; therefore, loading onto Ago2 [20] (and unpublished data), the core RNAi effector, is favored to achieve gene silencing.

2 Materials

2.1 Vector Modification

1. UASp vector [21].
2. Synthesized vector adaptors to yield UASp-KN, UASp-NB, and UASp-BX vectors, i.e., *Kpn*I/*Not*I-s and *Kpn*I/*Not*I-as adaptors to yield UASp-KN vector, *Not*I/*Bam*HI-s and *Not*I/*Bam*HI-as adaptors to yield UASp-NB vector, and *Bam*HI/*Xba*I-s and *Bam*HI/*Xba*I-as adaptors to yield UASp-BX vector (the adaptor sequences are shown in Table 1).
3. Restriction enzymes: *Kpn*I, *Not*I, *Bam*HI, and *Xba*I.
4. Agarose gel DNA isolation kit.
5. T4 DNA ligase.
6. Competent *E. coli* cells.
7. LB medium.
8. LB plates containing 100 ng/ml Ampicillin.
9. DNA preparation kit.

2.2 Artificial dme-miR-1-Mimicking siRNA Design and Expressing Vector Construction

1. Sense and antisense-strand oligos encoding artificial *dme-miR-1*-mimicking siRNAs. Each oligo is dissolved in ddH_2O to a final concentration of 100 μM.
2. 10× Taq polymerase buffer.
3. PCR instrument.
4. UASp-KN, UASp-NB, and UASp-BX vectors from Subheading 3.1.
5. Restriction enzymes: *Nhe*I and *Eco*RI.

Table 1
UASp vector adaptors

*Kpn*I/*Not*I-s	C ATCCCATATTCAGCC GCTAGC AGTCCACT GAATTC GGGCGAGACATCGGAG GC
*Kpn*I/*Not*I-as	GGCCGC CTCCGATGTCTCGCCC GAATTC AGTGGACT GCTAGC GGCTGAATATGGGAT GGTAC
*Not*I/*Bam*HI-s	GGCCGC ATCCCATATTCAGCC GCTAGC AGTCCACT GAATTC GGGCGAGACATCGGAG G
*Not*I/*Bam*HI-as	GATCC CTCCGATGTCTCGCCC GAATTC AGTGGACT GCTAGC GGCTGAATATGGGAT GC
*Bam*HI/*Xba*I-s	GATCC ATCCCATATTCAGCC GCTAGC AGTCCACT GAATTC GGGCGAGACATCGGAG T
*Bam*HI/*Xba*I-as	CTAGA CTCCGATGTCTCGCCC GAATTC AGTGGACT GCTAGC GGCTGAATATGGGAT G

Underlined bases are restriction enzyme sites, *Nhe*I or *Eco*RI

6. T4 DNA ligase.
7. Competent *E. coli* cells.
8. LB medium.
9. LB plates containing 100 ng/ml Ampicillin.
10. DNA preparation kit.
11. Restriction enzymes: *Not*I, *Bam*HI, and *Xba*I.

2.3 Transgene Engineering

1. Fly stocks: W^{1118}, sp/Cy0; *bam*Δ^{86}/TM3, sp/Cy0; *nosP-gal4vp16*/TM3.
2. Injection buffer: mix 20 μl of 100 mM PIPES, 15 μl of glycerol, 10 μl of 10 mM EDTA, and 55 μl of ddH_2O.
3. Egg laying plates containing 20 % sucrose, 20 % apple cider, and 2.4 % agar powder.
4. Micro-injection platform (Olympus).
5. Carl-Zeiss LSM7 laser confocal microscopy.

3 Methods

3.1 Vector Modification

For artificial *dme-miR-1*-mimicking siRNA cloning, three adaptors are individually inserted into original UASp vector [21] to construct three different modified UASp vectors, namely UASp-KN, UASp-NB, and UASp-BX. KN means that *Kpn*I and *Not*I sites are used as the adaptor insertion sites. Likewise, the adaptor insertion sites of UASp-NB and UASp-BX vector are *Not*I and *Bam*HI, and *Bam*HI and *Xba*I, respectively. Each adaptor contains *Nhe*I and *Eco*RI cloning sites for artificial *dme-miR-1*-mimicking siRNA insertion (Fig. 1).

1. To yield UASp-KN vector, digest ~1 μg of UASp vector with *Kpn*I and *Not*I, according to the manufacturer's instruction. To yield UASp-NB and UASp-BX vectors, *Kpn*I and *Not*I are substituted with *Not*I and *Bam*HI, and *Bam*HI and *Xba*I, respectively.
2. Mix 0.1 μg of *Kpn*I/*Not*I-s adapter with 0.1 μg of *Kpn*I/*Not*I-as adaptor and 2μl 10× Tag buffer in an Eppendorf tube and adjust the final volume to 20 μl by adding ddH_2O. Incubate the tube at 95 °C for 5 min and put it on bench to cool down to room temperature to Anneal a *Kpn*I and *Not*I double cohesive ends KN adaptor. Likewise, double-stranded cohesive ends NB and BX adaptors are produced by annealing *Not*I/*Bam*HI-s and *Not*I/*Bam*HI-as adaptors, and *Bam*HI/*Xba*I-s and *Bam*HI/*Xba*I-as adaptors, respectively. Resultant adaptor fragments are agarose gel-purified using a commercially available kit, according to the manufacturer's instruction.

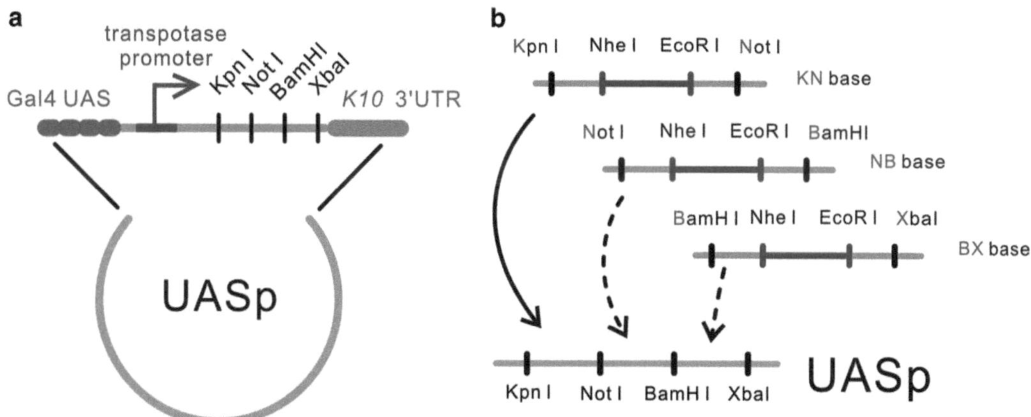

Fig. 1 Schematic diagram of vector modification. (**a**) The UASp vector, redrawn from Rørth (1998) [21]. (**b**) UASp vector modifications. Three independent adaptors bearing *Nhe*I and *Eco*RI were ligated into the UASp MCS and formed three different vectors: UASp-KN, UASp-NB, and UASp-BX. An artificial microRNA can then be integrated into the vector

3. Ligate gel-isolated double-stranded KN adaptor with UASp vector digested with *Kpn*I and *Not*I (from **step 1** in Subheading 3.1). To an Eppendorf tube, add 0.01 μg of digested vector, 0.05 μg of annealed adaptor, 1 μl of 10× T4 DNA buffer, 1 μl of T4 DNA ligase, and make the total volume to 10 μl. Gel-isolated NB and BX adaptors are ligated with UASp vector digested with *Not*I and *Bam*HI, and *Bam*HI and *Xba*I, respectively.
4. Transform *E. coli* with the reaction mixture and grow them on LB plates containing Ampicillin at 37 °C overnight.
5. Inoculate 2 ml of LB medium containing 100 ng/ml Ampicillin with each transformant and isolate plasmid DNA from *E. coli* using a DNA preparation kit according to the manufacturer's instruction.
6. Digest plasmid DNAs with *Nhe*I or *Eco*RI restriction enzymes, and sequence them, to confirm if modified vectors are obtained, as expected.

3.2 Artificial dme-miR-1-Mimicking siRNA Design and Expressing Vector Construction

Artificial *dme-miR-1*-mimicking siRNAs are designed through websites using siRNA design tools, such as Gene Link shRNA Explorer (http://www.genelink.com/sirna/shRNAi.asp) and DSIR (http://biodev.extra.cea.fr/DSIR/DSIR.html). An optimal siRNA sequence has a 40–60 % GC content, excludes repeats sequentially appearing more than three times, and is located within a region of the coding sequence that is within 50–100 nucleotides of the ATG start codon and within 50–100 nucleotides from the termination codon. Since the backbone of artificial siRNAs is *miR-1*, they should begin with a 5′-T to mimic the endogenous 5′ uracil of *miR-1* (*see* **Notes 1** and **2**). Selected siRNA sequences should also be Blast-searched against miRBase (http://www.mirbase.org/) and FlyBase (http://flybase.org/) to eliminate off-target effects (*see* **Note 3**). When necessary, it should be ensured that artificial siRNAs target all possible isoforms of their target genes (*see* **Note 4**).

Since miRNA-mimicking siRNAs mature from the hairpin-type pre-miRNA structure, flanking sequences [23] and mismatched bases are essential for siRNA maturation and should be included in the resultant artificial siRNAs to mimic the natural *miR-1*. We chose the same online tool used by Haley [15] (http://flybuzz.berkeley.edu/cgi-bin/constructhairpin.cgi). The pNE option was selected then the reverse complementary sequence of the siRNA designed above was entered into the "input sequence" textbox and the "create" button clicked. The resultant 71-nt oligos are then synthesized.

3.2.1 Single Hairpin Artificial dme-miR-1-Mimicking siRNA Construction

1. Anneal sense-strand and antisense-strand oligos by mixing 2 μl of 100 μM sense-strand oligo with 2 μl of 100 μM antisense-strand oligo, 1 μl of 10× Taq polymerase buffer in an Eppendorf tube and bring the final volume to 10 μl by adding ddH_2O. Then, incubate the mixture at 95 °C in a PCR instrument for 5 min, and then cool down to room temperature by following 20 min.
2. Dilute annealed oligo 10 times by adding 90 μl of ddH_2O.
3. Digest ~1 μg of UASp-KN vector with *Nhe*I and *Eco*RI, according to the manufacturer's instruction. UASp-NB and UASp-BX vectors can alternate UASp-KN vector.
4. Ligate an annealed oligo with UASp-KN vector digested with *Nhe*I and *Eco*RI to yield a plasmid to express *dme-miR-1*-mimicking siRNA, termed *UASp artmiR-X*. Different oligos can be inserted to other UASp-originating vectors, for instance, oligo 1 is inserted into UASp-KN, while oligo 2 is inserted into UASp-NB or UASp-BX vector. The construction is confirmed by PCR and/or sequencing (*see* **Note 5**).

3.2.2 Tandem Hairpin Construction

A single artificial *dme-miR-1*-mimicking siRNA may have a limited knockdown efficiency of the target gene; therefore, the use of two or three hairpins can ensure effective knockdown (*see* **Note 6**). Here, we show how to link three tandem hairpins.

1. Construct individually three independent single hairpin vectors as indicated in **step 2** in Subheading 3.1.
2. Excise the NB-hairpin fragment inserted into the UASp-NB vector by digesting the construct with *Not*I and *Bam*HI restriction enzymes (*see* **Note 7**).

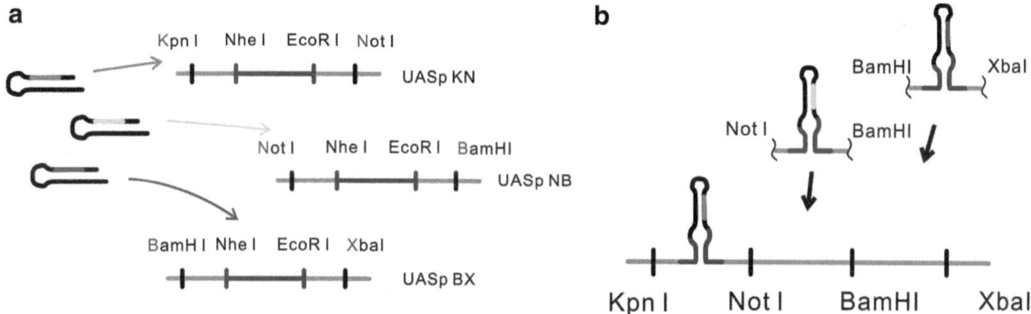

Fig. 2 Illustration of tandem hairpin creation. (**a**) By using appropriate restriction sites, different hairpins can be ligated into one vector. (**b**) Tandem hairpin procedure

3. Digest the UASp-KN vector containing the KN-hairpin fragment with *Not*I and *Bam*HI restriction enzymes.
4. Ligate ~100 bp NB-hairpin fragment into the linearized UASp-KN vector containing the KN-hairpin fragment. This is to link two hairpins in tandem in one vector. Insertion of the hairpin is confirmed by PCR and sequencing (*see* **Note 5**).
5. The third hairpin fragment bearing *Bam*HI and *Xba*I sites at 5′ and 3′ ends, respectively, can be inserted into the two hairpin-containing vector digested with *Bam*HI and *Xba*I (*see* Fig. 2).

3.3 Transgene Engineering

The micro-injection protocol is based on P-element [22] insertion.

1. Mix 2 μg/μl artificial *dme-miR-1*-mimicking siRNA expression plasmid and 1 μg/μl helper plasmid (pπ25.7 Δ2–3 wc) in Injection buffer. It is recommended that these plasmids are purified using a Qiagen Plasmid Midi Kit.
2. Collect fresh wild-type fly (w^{1118}) embryos (0–60 min) from an egg laying plate.
3. Inject the plasmid mix into the embryos via a micro-injection platform.
4. Incubate the injected embryos in a 25 °C growth cabinet for 36–48 h.
5. Transfer the larvae to fresh medium for continued growth.
6. Collect hatching adults and separate the sexes.
7. Cross them with mapping lines (we choose *bam*Δ^{86}) to obtain red eye offspring.
8. The insertions are then mapped and stable or balanced lines set up. The choice of crossing fly and balancers depends on individual preference.

9. To ensure the transgenes obtained are correct, the insert is amplified and sequenced.

3.4 Checking Phenotype for Knockdown Efficiency

The UAS/Gal4 system can be driven by different drivers and combined with the artificial *dme-miR-1*-mimicking siRNA; thus, theoretically, any gene can be knocked down in a tissue-specific manner. We tested the knockdown efficiency in the germline. *Bam* is a key factor controlling germline stem cell differentiation [24–26]; therefore, we generated a *UASp 3xart-bam* transgene line to silence *Bam* expression [27]. When knockdown was driven by *nosP*-gal4VP16 in the germline, the *UASp 3xart-bam* line phenocopied the *bam* mutant (immunohistochemistry procedure is described in [28]). The mRNA level, evaluated by real-time PCR, was highly reduced as expected. We subsequently targeted different genes and obtained similar knockdown effects [27]. These data indicate that two hairpins used together are capable of efficient gene silencing.

4 Notes

1. Since the synthesized oligos mimic the pre-*miR-1* processing, the intrinsic duplex structure should also mimic the natural *miR-1*. Thus, the thermodynamic properties of the duplex terminal are important for the sorting and strand selection by RISC. The mature miRNA duplex generally has thermodynamic asymmetry; the guide strand 5′ end has a lower thermodynamic energy than the passenger strand [29, 30]. Therefore, the siRNA 5′ T (T-A base pairing) accompanied with the *miR-1* 5′ U has a lower thermodynamic energy, which is compatible for loading onto RISC.
2. If the target CDS is too short to generate a siRNA, you can reduce the restrictions, such as using a wider GC content range or extending the target sequence to include the 3′ UTR.
3. To eliminating off-target effects, blast search the siRNA against annotated genes in FlyBase to exclude targeting other genes. MiRNA target recognition is dependent on its "seed" sequence (nucleotides 2–8 from the miRNA's 5′ end) [31]. Thus, exclude 21 nt siRNAs that have the same miRNA "seed" sequence by searching *Drosophila* miRNAs in mirBase.
4. For a rescue experiment, the 3′ UTR sequence should be the only target sequence, but ensure the *dme-miR-1*-mimicking siRNA can target all isoforms.
5. Inserted hairpin identification, PCR is a feasible method to identify the insertion(s). Synthesized oligos have cohesive termini and are ~100 bp long. We designed a pair of primers near the insertion site (see sequence below) to amplify the inserted hairpin(s). Blank UASp vector yields a PCR product of ~250 bp, thus one inserted hairpin yields a product

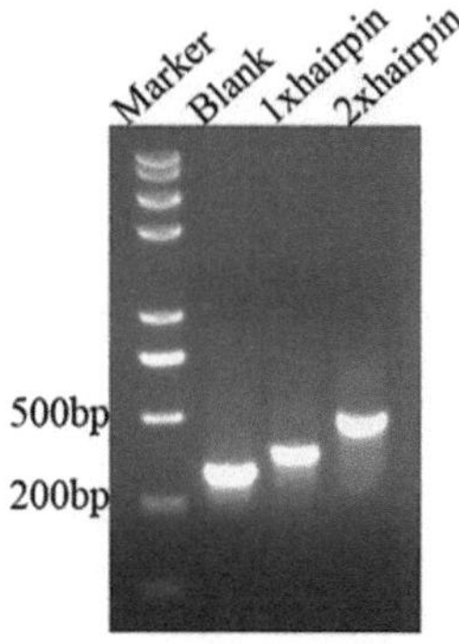

Fig. 3 PCR identification of insertions (1.5 % agarose gel). Addition of one hairpin yields PCR products that are shifted by ~100 bp in agarose gel electrophoresis

100 bp longer (*see* Fig. 3). Agarose gel electrophoresis of a 1.5 % agarose gel can identify vectors with different numbers of hairpins.

Primer/s: catttatgtaaacaataacgtgactgtgcg
Primer/as: accatgggtttaggtataatgttatcaagctc

6. Knockdown efficiency can be measured by phenotype, real-time PCR or Western blot. If the knockdown effect is not desirable, choose a different siRNA and link three hairpins.
7. In the tandem hairpin procedure, the digested fragments were too small to be detected; therefore, more than 20 μg of hairpin plasmid should be digested to ensure enough digestion fragment product.

References

1. Fire A, Xu S, Montgomery MK, Kostas SA, Driver SE, Mello CC (1998) Potent and specific genetic interference by double-stranded RNA in Caenorhabditis elegans. Nature 391(6669):806–811
2. Hammond SM, Bernstein E, Beach D, Hannon GJ (2000) An RNA-directed nuclease mediates post-transcriptional gene silencing in Drosophila cells. Nature 404(6775):293–296
3. Zamore PD, Tuschl T, Sharp PA, Bartel DP (2000) RNAi: double-stranded RNA directs the ATP-dependent cleavage of mRNA at 21 to 23 nucleotide intervals. Cell 101(1):25–33
4. Elbashir SM, Lendeckel W, Tuschl T (2001) RNA interference is mediated by 21-and 22-nucleotide RNAs. Gene Dev 15(2):188–200
5. Dickins RA, Hemann MT, Zilfou JT, Simpson DR, Ibarra I, Hannon GJ, Lowe SW (2005) Probing tumor phenotypes using stable and regulated synthetic microRNA precursors. Nat Genet 37(11):1289–1295
6. Silva JM, Li MZ, Chang K, Ge W, Golding MC, Rickles RJ, Siolas D, Hu G, Paddison PJ, Schlabach MR, Sheth N, Bradshaw J, Burchard J, Kulkarni A, Cavet G, Sachidanandam R, McCombie WR, Cleary MA, Elledge SJ, Hannon GJ (2005) Second-generation shRNA libraries covering the mouse and human genomes. Nat Genet 37(11):1281–1288
7. Alonso AGD, Gutierrez L, Fritsch C, Papp B, Beuchle D, Muller J (2007) A genetic screen identifies novel polycomb group genes in Drosophila. Genetics 176(4):2099–2108
8. Bokel C (2008) EMS screens: from mutagenesis to screening and mapping. Methods Mol Biol 420:119–138
9. Spradling AC, Stern D, Beaton A, Rhem EJ, Laverty T, Mozden N, Misra S, Rubin GM (1999) The Berkeley Drosophila Genome Project gene disruption project: single P-element insertions mutating 25% of vital Drosophila genes. Genetics 153(1):135–177

10. Oh SW, Kingsley T, Shin HH, Zheng Z, Chen HW, Chen X, Wang H, Ruan P, Moody M, Hou SX (2003) A P-element insertion screen identified mutations in 455 novel essential genes in Drosophila. Genetics 163(1): 195–201
11. Mathey-Prevot B, Perrimon N (2006) Drosophila genome-wide RNAi screens: are they delivering the promise? Cold Spring Harb Symp 71:141–148
12. Friedman A, Perrimon N (2004) Genome-wide high-throughput screens in functional genomics. Curr Opin Genet Dev 14(5):470–476
13. Juliano C, Wang J, Lin H (2011) Uniting germline and stem cells: the function of Piwi proteins and the piRNA pathway in diverse organisms. Ann Rev Genet 45:447–469
14. Siomi MC, Sato K, Pezic D, Aravin AA (2011) PIWI-interacting small RNAs: the vanguard of genome defence. Nat Rev Mol Cell Biol 12(4):246–258
15. Haley B, Hendrix D, Trang V, Levine M (2008) A simplified miRNA-based gene silencing method for Drosophila melanogaster. Dev Biol 321(2):482–490
16. Okamura K, Ishizuka A, Siomi H, Siomi MC (2004) Distinct roles for Argonaute proteins in small RNA-directed RNA cleavage pathways. Genes Dev 18(14):1655–1666
17. Kim VN, Han J, Siomi MC (2009) Biogenesis of small RNAs in animals. Nat Rev Mol Cell Biol 10(2):126–139
18. Ghildiyal M, Zamore PD (2009) Small silencing RNAs: an expanding universe. Nat Rev Genet 10(2):94–108
19. Tomari Y, Du T, Zamore PD (2007) Sorting of Drosophila small silencing RNAs. Cell 130(2):299–308
20. Ni JQ, Zhou R, Czech B, Liu LP, Holderbaum L, Yang-Zhou D, Shim HS, Tao R, Handler D, Karpowicz P, Binari R, Booker M, Brennecke J, Perkins LA, Hannon GJ, Perrimon N (2011) A genome-scale shRNA resource for transgenic RNAi in Drosophila. Nat Methods 8(5):405–407
21. Rorth P (1998) Gal4 in the Drosophila female germline. Mech Dev 78(1–2):113–118
22. Spradling AC, Rubin GM (1982) Transposition of cloned P elements into Drosophila germ line chromosomes. Science 218(4570): 341–347
23. Han J, Lee Y, Yeom KH, Nam JW, Heo I, Rhee JK, Sohn SY, Cho Y, Zhang BT, Kim VN (2006) Molecular basis for the recognition of primary microRNAs by the Drosha-DGCR8 complex. Cell 125(5):887–901
24. McKearin DM, Spradling AC (1990) bag-of-marbles: a Drosophila gene required to initiate both male and female gametogenesis. Genes Dev 4:2242–2251
25. Chen D, McKearin DM (2003) A discrete transcriptional silencer in the bam gene determines asymmetric division of the Drosophila germline stem cell. Development 130:1159–1170
26. Song X, Wong MD, Kawase E, Xi R, Ding BC, McCarthy JJ, Xie T (2004) Bmp signals from niche cells directly repress transcription of a differentiation-promoting gene, bag of marbles, in germline stem cells in the Drosophila ovary. Development 131(6):1353–1364
27. Wang H, Mu Y, Chen D (2011) Effective gene silencing in Drosophila ovarian germline by artificial microRNAs. Cell Res 21(4):700–703
28. Xia L, Jia S, Huang S, Wang H, Zhu Y, Mu Y, Kan L, Zheng W, Wu D, Li X, Sun Q, Meng A, Chen D (2010) The Fused/Smurf complex controls the fate of Drosophila germline stem cells by generating a gradient BMP response. Cell 143(6):978–990
29. Czech B, Zhou R, Erlich Y, Brennecke J, Binari R, Villalta C, Gordon A, Perrimon N, Hannon GJ (2009) Hierarchical rules for Argonaute loading in Drosophila. Mol Cell 36(3):445–456
30. Ghildiyal M, Xu J, Seitz H, Weng Z, Zamore PD (2010) Sorting of Drosophila small silencing RNAs partitions microRNA* strands into the RNA interference pathway. RNA 16(1):43–56
31. Bartel DP (2009) MicroRNAs: target recognition and regulatory functions. Cell 136(2):215–233

Chapter 19

Isolation and Bioinformatic Analyses of Small RNAs Interacting with Germ Cell-Specific Argonaute in Rice

Reina Komiya and Ken-Ichi Nonomura

Abstract

The small noncoding RNAs in plants are categorized into two major classes, 21-nucleotides (nt) micro RNA (miRNA) and 21- or 24-nt small-interfering RNA (siRNA). ARGONAUTE (AGO) proteins associate with small RNAs and play central roles in transcriptional and posttranscriptional gene regulation. In plants, AGO1–miRNA complexes mainly regulate developmental processes, and AGO4–siRNA complexes suppress the activity of transposons and exogenous viral infections via RNA-directed DNA methylation. In many animal species, the PIWI-subfamily AGOs interact with PIWI-interacting RNAs (piRNAs), which are most commonly 24–34 nt, and function to tame transposons and to regulate mRNA translation and stability in the germline. The rice protein MEIOSIS ARRESTED AT LEPTOTENE1 (MEL1) is a plant AGO member that has roles specific to development and maintenance of germ cells before meiosis. MEL1-binding small RNAs are mainly 21 nt, have a 5′-terminal cytosine, and are distinct from animal piRNAs. In this chapter, we describe methods for RNA-immunoprecipitation (RNA-IP) using a specific antibody that recognizes MEL1 and subsequent purification of MEL1-associating small RNAs from the IP fraction. We also introduce the bioinformatic procedures including mapping, annotation, and identifying small RNA clusters on the rice genome.

Key words Small RNA, ARGONAUTE, Small RNA-immunoprecipitation, Germ cell, Rice

1 Introduction

Sexual reproduction of land plants is distinct from that of animals. In most animals, meiosis directly produces haploid gametes, and meiosis directly produces haploid gametes. In contrast, a land plant has two different plant bodies, sporophyte and gametophyte. In angiosperms, diploid sporophytes produce haploid spores by meiosis. The individual spores then germinate, undergo mitosis, and produce multicellular haploid gametophytes, the pollen and embryo sac that contain gametes. This unique life cycle is called the "alteration of generations" [1].

Recently, we discovered a rice gene, designated *MEIOSIS ARRESTED AT LEPTOTENE1* (*MEL1*) that encodes a protein in the ARGONAUTE (AGO) family [2]. *MEL1* expression is prominent in

Mikiko C. Siomi (ed.), *PIWI-Interacting RNAs: Methods and Protocols*, Methods in Molecular Biology, vol. 1093, DOI 10.1007/978-1-62703-694-8_19, © Springer Science+Business Media, LLC 2014

the sporophytic germ cells before meiosis and is downregulated rapidly during the transition to meiosis. The *mel1* mutation causes aberrant vacuolation within premiotic germ cells, meiotic arrest at the leptotene stage, and absence of fertile gametes [2]. MEL1 is one of a few plant AGOs known to have its specific roles in development and maintenance of germ cells before meiosis.

Plant AGOs are categorized into several subgroups, AGO1, AGO4, AGO7, and MEL1 [3]. AGO1 protein mainly binds to 21-nucleotides (nt) microRNAs (miRNA), which are produced from the stem loop structure of a primary miRNA transcript. AGO1 protein cleaves target mRNAs in a sequence-dependent manner via the slicer activity of PIWI domain [4, 5]. AGO4 associates with 24-nt small interfering RNAs (siRNAs) and causes gene silencing via DNA methylation [6]. AGO7 is required for the *trans*-acting siRNA (tasiRNA) pathway, which is triggered by AGO1-dependent cleavage of single-stranded RNAs, followed by RNA-DEPENDENT RNA POLYMERASE6 (RDR6)-dependent double-strand formation and DICER LIKE4 (DCL4)-dependent processing to create 21-nt tasiRNAs [7]. AGO7/1-tasiRNA complexes suppress target genes in *trans*.

MEL1 is a member of the MEL1 clade, which is divergent and functionally unknown of the AGO subgroups. Based on RNA-immunoprecipitation analyses in rice, most small RNAs that bind to MEL1 are 21-nt siRNAs with a 5′-terminal cytosine that are derived from intergenic and non-repetitive regions of the genome (Komiya et al. in preparation). Moreover, few, if any, targets of MEL1–small RNA complexes have been identified. In animals, the germline-enriched PIWI-interacting RNAs (piRNAs) are usually 24–34 nt in length and are a major type of small noncoding RNAs [8, 9]. These findings clearly indicate that during germline development, abundant expression of small RNAs is common, but small RNA pathways that function in those stages are distinct, between animals and plants.

In this chapter, we describe methods for purifying the MEL1–small RNA complex via immunoprecipitation (IP) with an anti-MEL1 antibody, for isolating MEL1-binding small RNAs from the resulting immunoprecipitates, and for conducting deep-sequencing and bioinformatic analyses of these small RNAs. Though the protocol in this chapter is adjusted to MEL1 AGO expressed specifically in young florescence of rice, it may be applicable to other species and tissues with some slight modifications.

2 Materials

Use RNase-free water to prepare all solutions, except those used for electrophoresis. Wearing gloves is recommended to limit contamination of reagents with RNase.

2.1 Materials for Preparation of Total Small RNAs from Reproductive Tissues of Rice

1. Fresh young rice inflorescences (panicles): Carefully extract 3 cm young panicles by clean forceps from the stem at the bolting stage (premeiotic stage), about 15 days before heading of panicles. Immediately after harvesting, freeze the panicles with liquid nitrogen, and store at −80 °C.
2. TRIZOL® Reagent (Invitrogen); Guanidium thiocyanate-based RNA extraction reagents.
3. mirVana™ miRNA isolation kit (Ambion) for enrichment of small RNAs.
4. RNase-free water (e.g., DEPC-treated water).
5. 70 % EtOH (in RNase-free water).

2.2 Components of Reagents for Small RNA-IP

1. Total protein extraction buffer: 150 mM NaCl, 50 mM Tris–HCl (pH 7.0), 0.1 % Tween 20, 10 % glycerol, 1 mM Dithiothreitol (DTT), 1 mM Pefabloc SC (Roche), 1× Complete Protease Inhibitor Cocktail (Roche; modified [10]).
2. Wash buffer: 20 mM Tris–HCl (pH 7.5), 300 mM NaCl, 5 mM $MgCl_2$, 5 mM DTT, 0.1 % NP-40 (SIGMA), 1× Complete Protease Inhibitor [11, 12].
3. 2× Sodium dodecyl sulfate (SDS) sample buffer: 0.1 M Tris–HCl (pH 6.8), 12 % 2-mercaptoethanol, 4 % SDS, 20 % glycerol, Bromophenol Blue.
4. Protein G-Sepharose 4 Fast Flow (GE Healthcare).
5. MULTIGEL II Mini 4/20 (13 W) (Cosmo Bio).
6. Running buffer: 0.05 M Tris (hydroxymethyl) aminomethane, 0.4 M Glycine, 0.1 % SDS.
7. Silver Stain MS kit (Wako).
8. ISOGEN-LS buffer (Nippongene).
9. T4 polynucleotide kinase (PNK) (Nippongene).
10. [γ-^{32}P]ATP (7,000 Ci/mmol) (HAS).
11. GE-25 columns (GE Healthcare).
12. Gel Loading Buffer II (Ambion).
13. 15 % Acrylamide/6 M Urea Gel: 1×TBE (Tris-borate, ethylenediaminetetraacetic acid (EDTA)), 15 % Acrylamide (Acrylamide:Bis = 19:1), 6 M Urea, TEMED (*N*,*N*,*N*′,*N*′-Tetramethyl-ethylenediamine), 0.1 % APS (Ammonium persulfate solution) (*see* **Note 1**).
14. 10×TBE stock buffer: 0.89 M Tris, 0.89 M Boric acid, 0.02 M EDTA (pH 8.0).

3 Methods

This section consists of three parts: (1) Extraction of total small RNAs, (2) Small RNA-IP, and (3) Sequencing and bioinformatic analysis of small RNAs. The methods described in both Subheadings 3.1 and 3.2 are performed on ice or at 4 °C.

3.1 Extraction of Low-Molecular-Weight RNAs Expressed During Premeiotic Stages

To extract high-quality samples of total small RNAs, two steps of RNA purification are recommended. First, samples of total RNA are isolated via the guanidinium thiocyanate phenol chloroform method (GTC). Next, the mirVana™ miRNA isolation kit is used to purify low-molecular-weight RNAs (<200 nt) from samples of total RNA.

Extraction of Total RNAs

1. Use stainless beads and the SHAKE MASTER (Bio Medical Science) to grind 50–100 mg of young rice panicles (3 cm) in liquid nitrogen. Add 1 ml TRIZOL and mix thoroughly; then incubate the mixture for 5 min at room temperature. Next, add 0.2 ml chloroform, vortex for 15 s, and incubate for 3 min at room temperature. Centrifuge the mixture at 12,000 × *g* for 15 min at 4 °C, and transfer the aqueous phase to a new microtube.
2. Add 0.5 ml isopropanol, mix thoroughly, and incubate for 10 min at room temperature. Centrifuge at 12,000 × *g* for 10 min at 4 °C, and carefully remove supernatant.
3. Wash the pellet twice with 1 ml 70 % EtOH.
4. Dry and dissolve the pellet containing total RNA in 30–50 μl RNase-free water (*see* **Note 2**).

Isolation of Small RNAs from Total RNAs

5. Mix 50–100 μg of total RNA with 5 volumes of Lysis/Binding buffer in the mirVana kit. Add 1/10 volume of miRNA Homogenate Additive from the kit to the mixture, mix thoroughly, and incubate for 10 min on ice.
6. Add 1/3 volume of 100 % EtOH, apply the mixture onto the filter cartridge in the kit, centrifuge at 5,000 × *g* for 1 min, and collect the filtrate in a new microtube.
7. Add 2/3 volume of 100 % EtOH to the filtrate, apply the mixture onto the second filter cartridge, centrifuge at 5,000 × *g* for 1 min, and discard the flow-through.
8. Apply 700 μl miRNA Wash Solution 1 to the cartridge, centrifuge at 5,000 × *g* for 1 min, and discard the flow-through. Wash the filter cartridge twice more; use 500 μl of Wash Solution 2/3 and a 1-min centrifugation step for each of these washes.

9. Transfer the cartridge onto a fresh collection tube, apply 50 μl of 95 °C Elution Solution to the cartridge, and incubate for 2 min at room temperature. Collect the eluate, which should contain purified small RNAs, by centrifugation at 10,000×*g* for 1 min, and store it at −20 °C or colder.

3.2 Immunoprecipitation of MEL1–Small RNA Complexes

The sample volume, the quality of the antibody, washing time, and buffer composition are each critical for successful RNA-IP.

Small RNA-IP

1. Grind 200 mg of rice young panicles (3 cm) in liquid nitrogen in a sterilized mortar and pestle. Add extraction buffer (1–1.6 ml buffer per 200 mg of tissues) and mix gently. Remove the debris by centrifugation at 5,000×*g* for 10 min at 4 °C, and transfer the supernatant to a fresh tube. Next, recentrifuge at 20,000×*g* for 10 min at 4 °C. Transfer the supernatant to a new tube (*see* **Note 3**).
2. Add 50 μl of protein G-sepharose suspension to 1 ml of each sample for pre-cleaning and incubate these mixtures for 3 h at 4 °C on a rotator. Remove G-sepharose by centrifugation at 12,000×*g* for 20 s at 4 °C. Transfer the supernatant to a new microtube.
3. Add 1–5 μg of an anti-MEL1 antibody to the sample and incubate for 1 h at 4 °C on a rotator. In addition, preimmune serum is used as a negative control (*see* **Note 4**).
4. Add 50 μl of protein G-sepharose suspension to the mixture and incubate overnight at 4 °C on a rotator. Recover G-sepharose by centrifugation at 12,000×*g* for 20 s at 4 °C and carefully remove the supernatant.
5. Add 1 ml of wash buffer and incubate for 20 min at 4 °C on a rotator. Centrifuge the sample at 12,000×*g* for 20 s at 4 °C and remove supernatant. Repeat this wash step three more times (*see* **Note 5**).
6. Add 120 μl of wash buffer, mix gently, and use 20 μl suspension for electrophoresis of MEL1 complexes (go to **step 7** in Subheading 3.2) and the remaining 100 μl for elution of small RNAs from MEL1 proteins (go to **step 9** in Subheading 3.2).

Electrophoresis of the MEL1 Complexes

7. Add an equal volume of 2×SDS sample buffer to the 20 μl suspension (**step 6** in Subheading 3.2), incubate this mixture at 95 °C for 5 min, then transfer the sample tube to ice. Load the samples on a MULTIGEL II Mini 4/20 and separate the proteins via electrophoresis at 25 mA.
8. Use the Silver Stain MS kit to stain the proteins in the gel (Fig. 1a) (*see* **Note 6**).

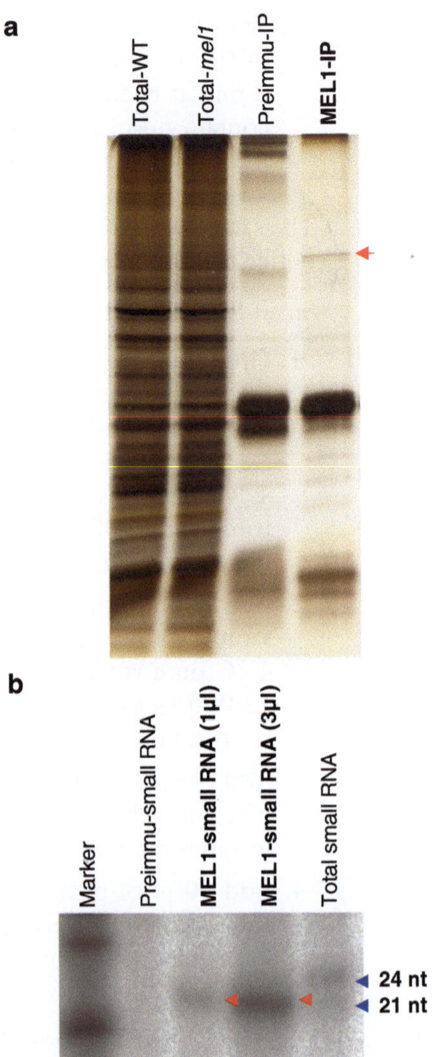

Fig. 1 Purification of MEL1 complexes and small RNAs. (**a**) Peptide-specific antibodies and preimmunoserum were used to immunoprecipitate MEL1 complexes; these purified complexes and total protein (WT and *mel1* mutants) at rice reproductive stages were separated on 10 % SDS-PAGE and visualized by silver staining. The MEL1 protein band is indicated by a *red arrow*. (**b**) Small RNAs were extracted from MEL1 complexes and visualized by RI. MEL1–small RNA were detected mainly at the size of 21 nucleotides (*lane 3*: 1 μl of MEL1–small RNA (**step 12** in Subheading 3.2), *lane 4*: 3 μl of MEL1–small RNA). Small RNAs purified from total RNA were used as the positive control (*lane 2*) and small RNAs extracted from preimmune antisera were used as the negative control (*lane 5*). *Lane 1* shows the marker including an upper band of 30 bp and a lower band of 20 bp. *Red* and *blue arrowheads* show 21nt-MEL1–small RNAs, and 21 and 24 nt total small RNAs

Extraction of Small RNAs from the IP Fraction

9. Add 300 μl of ISOGEN-LS buffer to the 100 μl portion of the suspension that was not used for electrophoretic analysis (**step 6** in Subheading 3.2), vortex the tube, and incubate for 5 min at room temperature. Next, add 80 μl of chloroform to this sample, vortex, and incubate for 3 min at room temperature. Remove the protein G-sepharose by centrifugation at 12,000 × *g* for 15 min at 4 °C, and transfer the aqueous phase to a new microtube.
10. Add 0.5 μl glycogen (20 μg/μl) as a carrier and 0.8 volume of isopropanol to the aqueous phase; incubate this mixture for 1 h at −80 °C. Recover the small RNA-pellet by centrifugation at 12,000 × *g* for 10 min at 4 °C, and then remove supernatant.
11. Wash the pellet with 70 % EtOH, dry the pellet, and dissolve it in 10 μl RNase-free water.

Validation of Small-RNA Extraction by [γ-^{32}P]ATP-End Labeling

12. Transfer 1–3 μl MEL1–small RNAs samples (**step 11** in Subheading 3.2) and 10–50 ng total small RNA (**step 9** in Subheading 3.1) into a new tube. Add 0.6 μl of T4 polynucleotide kinase (PNK), 2 μl [γ-^{32}P]ATP (7,000 Ci/mmol), and 3 μl of 5×PNK buffer to the tube; the final reaction volume will be 15 μl. Incubate this reaction mixture at 37 °C for 1 h, transfer it onto a GE-25 column to remove the free [γ-^{32}P] ATP, and collect the flow-through which should contain labeled small RNAs in a new microtube (*see* **Note 7**).
13. Add an equal volume of Gel Loading Buffer II to the flow-through sample, and mix thoroughly. Heat the sample for 10 min at 65 °C, and incubate it on the ice at least for 5 min. Load the sample onto a 15 % acrylamide/6 M urea gel well and separate the components via electrophoresis at 150 V at 4 °C. Use the FLA9000 (GE Healthcare) imaging system to visualize the labeled small RNAs (Fig. 1b).

3.3 Deep-Sequencing and Classification of Small RNAs that Bind AGO Proteins

This section describes the bioinformatic procedures to classify deep-sequencing data of small RNAs that bind AGO proteins. Small RNA sequences are cloned and sequenced by deep-sequencing. We describe bioinformatic analyses of mapping, annotation, classification, and clustering performed by mapping tools and in-house program. An overview of the bioinformatic classification of small RNAs is shown in Fig. 2.

1. Use the standard Illumina protocol to produce two small RNA libraries; one library should represent the total small RNA sample (Subheading 3.1) and the other library should represent the MEL1–small RNA sample (Subheading 3.2). Sequence each clone within these libraries (*see* **Note 8**).

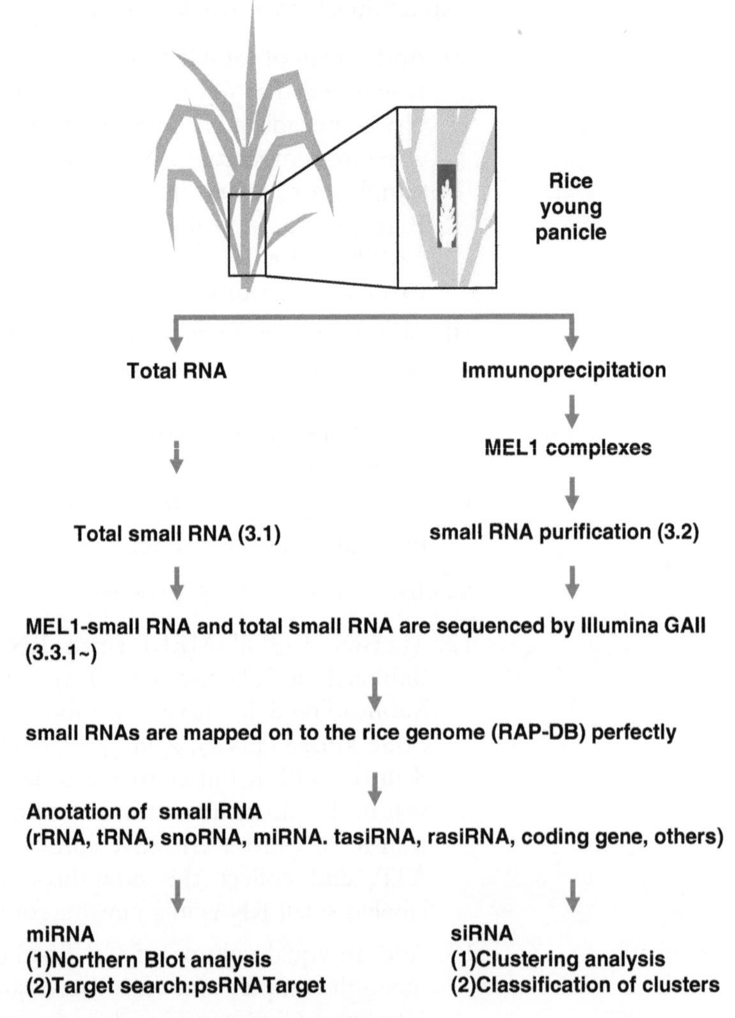

Fig. 2 Flow chart of small RNA classification

2. After removing adaptor sequences, small RNA reads between 18- and 28-nt lengths are extracted and mapped on the rice genome (*see* **Note 9**).
3. Classify each sequence, total small RNAs and MEL1–small RNAs, into one of several appropriate subgroups (rRNA, tRNA, snoRNA, miRNA, tasiRNA, nat-siRNA, rasiRNA, and coding gene) according to the sequence information from public databases and publications (*see* **Note 10**).
4. Use the small-RNA clustering method to identify genomic regions that generate small RNAs. We used all small RNAs other than rRNAs, tRNAs, and snoRNAs in this clustering analysis. Regions that meet three criteria are defined as a cluster in our analysis (modified [5, 13]); these criteria are as follows: (1) two neighboring small RNAs within an interval of

200 nt or less are gathered into a same cluster, (2) one cluster contains at least ten unique small RNAs, or (3) the total count of small-RNA reads is over 1,000 in a cluster (*see* **Note 11**).

5. Classify small RNA clusters into several categories such as miRNAs, rasiRNAs, or tasiRNAs.

4 Notes

1. Wipe the glass plates and comb well twice with each alcohol solution, 100 % isopropanol and then with 70 % EtOH. The 10 ml mixture of 15 % acrylamide/6 M urea in RNase-free water should be rotated for 15 min at room temperature. After deaeration, keep the solution on ice and add 13 μl TEMED and 100 μl 10 % APS; mix the solution immediately, pour the mixture into the mini-gel plates, and incubate for 1 h at room temperature.
2. 30–50 μg of total RNAs are extracted from about 50–100 mg young panicles. A rice panicle of 3 cm long is about 50 mg.
3. In addition to validation of small RNA extraction and construction of small RNA libraries, we also use the IP fraction for mass spectrometry to confirm that the IP was successful. Therefore, we usually use 1,000 mg of 3 cm panicles all at once to increase the yield of MEL1 complexes. The ground sample is collected in a 50 ml conical tube, and 7.5 ml extraction buffer are added. In the next step (**step 2** in Subheading 3.2), the supernatant is divided into several microtubes (800–1000 μl per tube).
4. The epitopes were carefully determined to avoid cross-reaction with the other AGOs. In our case, two synthetic peptides of 16 or 17 amino acids were conjugated to the appropriate carrier, and the mixture of two peptides were used to immunize a rabbit. The antisera were affinity purified.
5. We wash MEL1–small RNA complexes four times. More extensive washing may result in higher quality samples of AGO-small RNAs.
6. We checked the MEL1 IP fractions by mass spectrometry to avoid cross-reaction of antibody before small RNA extraction. After washing the G-sepharose (**step 5** in Subheading 3.2), add 30 μl of 1×SDS sample buffer to the sepharose pellet, mix, incubate at 95 °C for 5 min, cool on ice, load the samples on MULTIGEL II Mini 4/20, and separate the components by electrophoresis. Cut out the expected band from a stained gel, treat with trypsin, recover the peptide samples, and evaluate these peptides by mass spectrometry, HCT ultra (BRUKER), according to the manufacturer's instruction. In all procedures,

using disposal knife, wearing gloves, masks and hair-masks is effective for avoiding contamination with keratin. Running buffer (1×TBE) for mass spectrometry is filtered with Stericup & Steritop (MILLIPORE).

7. Because the yield of MEL1–small RNAs is usually very low, we do not estimate RNA concentration. The quality of MEL1–small RNAs is checked via electrophoresis of [γ-^{32}P]-labeled small RNAs directly.
8. The quality of the samples of total small RNA should be checked via urea-polyacrylamide gel electrophoresis or a bio-anlyzer (Agilent Technologies) before cloning for deep sequencing. MEL1–small RNAs are used directly for cloning after evaluation them via electrophoresis of [γ-^{32}P]-labeled small RNAs. When cloned, MEL1–small RNAs are amplified by PCR (15 cycles). After the gel electrophoresis, gel slices that contain small RNAs of 15–30 nt are cut out based on size marker; these gel slices are then used for elution and deep-sequencing of small RNAs. We used the Illumina Genome Analyzer II for small RNAs-sequence.
9. The rice genome information (IRGSP build 5.0) is available at the online at

   RAP-DB (http://rapdblegacy.dna.affrc.go.jp/document/index.html). We used Short Oligonucleotide Analysis Package, SOAP (http://soap.genomics.org.cn/) for mapping small RNA-seq reads onto the rice genome. Only the small RNAs that matched perfectly with genomic sequences are used for further analyses.
10. Public databases or publications used in this study are as follows:

    RAP-DB (http://rapdblegacy.dna.affrc.go.jp/document/index.html) for coding genes,

    Rfam (http://rfam.sanger.ac.uk/) for snoRNAs, the information in Zhu et al. [14] for tasiRNAs, and the information in Osato et al. [15] for nat-siRNAs. Small RNAs were classified using SOAP. The information of rice repeats of rRNAs, tRNAs, and pre-miRNAs were kindly provided by the RAP-DB administrator. The other repeats were extracted from publicly available genome sequence (RAP-DB build 5.0), in which the repeats identified by the repeat masker were shown in small letters.
11. The total count is the sum of [read number of small RNA/genome hit number] in a cluster. The genome hit number indicates how many sequences identical to the small RNA appear in the genome.

    Here, we mainly introduce a procedure of the cluster analysis, because the majority of MEL1–small RNAs are classified as siRNAs in procedure 3.3, **step 3**. If your interest is in miRNA candidates, analyze your candidates online at the plant small RNA target analysis server, psRNAtarget (http://plantgrn.noble.org/psRNATarget/), and then use the modified RACE method [16] to examine the actual target cleavage sites.

Acknowledgments

This work was supported by JSPS KAKENHI Grant No. 21678001 and No. 25252004 (to K.I.N.) and JSPS Research Fellowships for Young Scientists (to R.K.).

References

1. Graham EL (1985) The origin of the life cycle of land plants. Am Sci 73:178–186
2. Nonomura K, Morohoshi A, Nakano M, Eiguchi M, Miyao A, Hirochika H et al (2007) A germ cell specific gene of the ARGONAUTE family is essential for the progression of premeiotic mitosis and meiosis during sporogenesis in rice. Plant Cell 19:2583–2594
3. Kapoor M, Arora R, Lama T, Nijhawan A, Khurana JP, Tyagi AK et al (2008) Genome-wide identification, organization and phylogenetic analysis of Dicer-like, Argonaute and RNA-dependent RNA Polymerase gene families and their expression analysis during reproductive development and stress in rice. BMC Genomics 9:451–467
4. Mallory AC, Elmayan T, Vaucheret H (2008) MicroRNA maturation and action-the expanding roles of ARGONAUTEs. Curr Opin Plant Biol 11:560–566
5. Wu L, Zhang Q, Zhou H, Ni F, Wu X, Qi Y (2009) Rice MicroRNA effector complexes and targets. Plant Cell 21:3421–3435
6. Matzke M, Kanno T, Daxinger L, Huettel B, Matzke AJ (2009) RNA-mediated chromatin-based silencing in plants. Curr Opin Cell Biol 21:367–376
7. Allen E, Howell MD (2010) miRNAs in the biogenesis of trans-acting siRNAs in higher plants. Semin Cell Dev Biol 21:798–804
8. Saxe JP, Lin H (2011) Small noncoding RNAs in the germline. In: Sassone-Corsi P, Fuller MT, Braun R (eds) Germ cells. Cold Spring Harbor Laboratory, New York, pp 131–146
9. Siomi MC, Sato K, Pezic D, Aravin AA (2011) PIWI-interacting small RNAs: the vanguard of genome defence. Nat Rev Mol Cell Biol 12:246–258
10. Mathieu J, Warthmann N, Kuttner F, Schmid M (2007) Export of FT protein from phloem companion cells is sufficient for floral induction in *Arabidopsis*. Curr Biol 17:1055–1060
11. Takeda A, Iwasaki S, Watanabe T, Utsumi M, Watanabe Y (2008) The mechanism selecting the guide strand from small RNA duplexes is different among argonaute proteins. Plant Cell Physiol 49:493–500
12. Watanabe T, Totoki Y, Toyoda A, Kaneda M, Kuramochi-Miyagawa S, Obata Y et al (2008) Endogenous siRNAs from naturally formed dsRNAs regulate transcripts in mouse oocytes. Nature 453:539–543
13. Johnson C, Kasprzewska A, Tennessen K, Fernandes J, Nan GL, Walbot V et al (2009) Clusters and superclusters of phased small RNAs in the developing inflorescence of rice. Genome Res 19:1429–1440
14. Zhu QH, Spriggs A, Matthew L, Fan L, Kennedy G, Gubler F et al (2008) A diverse set of microRNAs and microRNA-like small RNAs in developing rice grains. Genome Res 18:1456–1465
15. Osato N, Yamada H, Satoh K, Ooka H, Yamamoto M, Suzuki K et al (2003) Antisense transcripts with rice full-length cDNAs. Genome Biol 5:R5
16. Llave C, Xie Z, Kasschau KD, Carrington JC (2002) Cleavage of Scarecrow-like mRNA targets directed by a class of *Arabidopsis* miRNA. Science 297:2053–2056

ERRATA

DNA Methylation in Mouse Testes

Satomi Kuramochi-Miyagawa, Kanako Kita-Kojima, Yusuke Shiromoto, Daisuke Ito, Hirotaka Koshima, and Toru Nakano

Mikiko C. Siomi (ed.), *PIWI-Interacting RNAs: Methods and Protocols*, Methods in Molecular Biology, vol. 1093, DOI 10.1007/978-1-62703-694-8_8, © Springer Science+Business Media, LLC 2014

DOI 10.1007/978-1-62703-694-8_20

The publisher regrets that in Chapter 8, Figure 2 was incorrect. The correct version is:

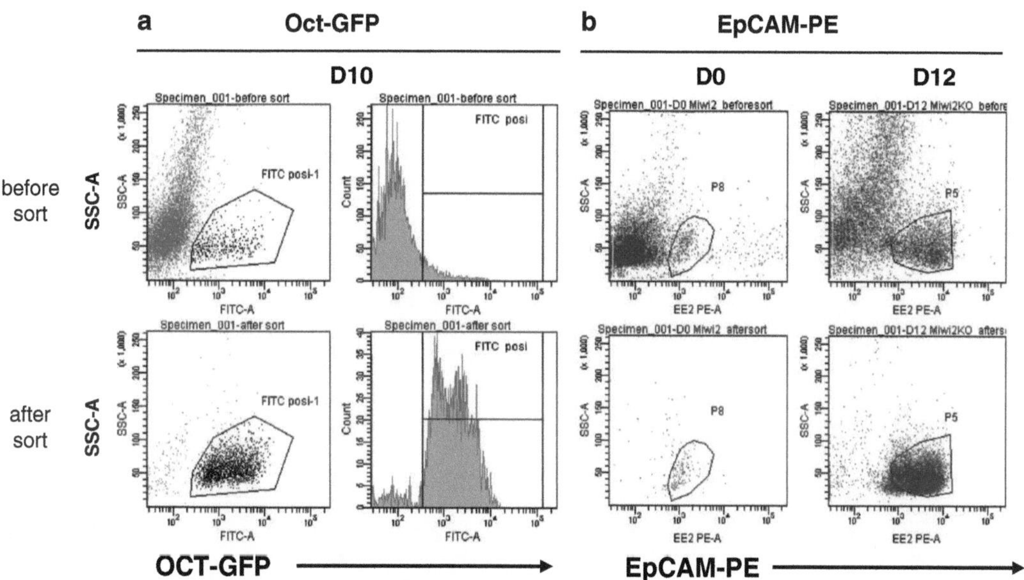

Fig. 2 Flow cytometry analysis before and after sorting. (a) The Oct-GFP (FITC) positive germ cells were isolated from the day 10 testis after birth of Oct-GFP mouse by FACSAria (BD). (b) The PE-EpCAM positive germ cells were isolated from the day 1 or day 12 testes after birth by FACSAria (BD)

The online version of the original chapter can be found at http://dx.doi.org/10.1007/978-1-62703-694-8_8

Analysis of Piwi-Loaded Small RNAs in Terahymena

Mikiko C. Siomi

Mikiko C. Siomi (ed.), *PIWI-Interacting RNAs: Methods and Protocols*, Methods in Molecular Biology, vol. 1093,
DOI 10.1007/978-1-62703-694-8_17, © Springer Science+Business Media, LLC 2014

DOI 10.1007/978-1-62703-694-8_20

The publisher regrets that in the online and print versions of this book, the title of Chapter 17 reads incorrectly in both the front matter and on the chapter opening page.

This title should read "Analysis of Piwi-Loaded Small RNAs in *Tetrahymena*"

The online version of the original chapter can be found at
http://dx.doi.org/10.1007/978-1-62703-694-8_17

Index

Mikiko C. Siomi (ed.), *PIWI-Interacting RNAs: Methods and Protocols*, Methods in Molecular Biology, vol. 1093, DOI 10.1007/978-1-62703-694-8, © Springer Science+Business Media, LLC 2014

MIX
Papier aus verantwortungsvollen Quellen
Paper from responsible sources
FSC® C105338

If you have any concerns about our products,
you can contact us on
ProductSafety@springernature.com

In case Publisher is established outside the EU,
the EU authorized representative is:
Springer Nature Customer Service Center GmbH
Europaplatz 3, 69115 Heidelberg, Germany

Printed by Libri Plureos GmbH
in Hamburg, Germany